Introducing
COGNITIVE BEHAVIORAL THERAPY

By

Daniel Anderson

COPYRIGHT © 2019.

ALL RIGHTS RESERVED.

No part of this publication may be reproduced, distributed, or transmitted in any form or by any means, including photocopying, recording, or other electronic or mechanical methods, or by any information storage and retrieval system without the prior written permission of the publisher, except in the case of very brief quotations embodied in critical reviews and certain other noncommercial uses permitted by copyright law.

Table of Contents

TAKING COGNIZANCE OF THE COMMON MENTAL HEALTH ISSUES 4

BECOME FAMILIAR WITH THE BASIC PRINCIPLES OF CBT AND UNDERSTAND HOW IT WORKS 35

HOW TO DEFINE A SPECIFIC GOAL TO WORK TOWARD OVER THE COURSE OF 6 WEEKS 67

CBT FAQS 79

ALTERNATIVE COGNITIVE BEHAVIOURAL APPROACHES 93

APPLYING COGNITIVE BEHAVIORAL THERAPY (CBT) TO OVER-THINK NEGATIVE PATTERNS 105

MOST COMMON ERRORS MADE IN COGNITIVE-BEHAVIORAL THERAPY 118

CHAPTER ONE

TAKING COGNIZANCE OF THE COMMON MENTAL HEALTH ISSUES

Anxiety

Feeling worried or nervous is a normal part of everyday life. Everyone frets or feels anxious from time to time. Mild to moderate anxiety can help you focus your attention, energy, and motivation. If anxiety is severe, you may have feelings of helplessness, confusion, and extreme worry that are out of proportion with the actual seriousness or likelihood of the feared event. Overwhelming anxiety that interferes with daily life is not normal. This type of anxiety may be a symptom of generalized anxiety disorder, or it may be a symptom of another problem, such as depression.

Jane's story

"Jane has always been a worrier, but it never interfered with her life before. Lately, however, she's been feeling keyed up all the time. She's paralyzed by an omnipresent sense of dread, and worries constantly about the future. Her worries make it difficult to concentrate at work, and when she gets home she can't relax. Jane is also having sleep difficulties, tossing and turning for hours before she falls asleep. She also gets frequent stomach cramps and diarrhea, and has a chronic stiff neck from muscle tension. Jane feels like she's on the verge of a nervous breakdown."

What is GAD?

Everyone gets worried sometimes, but if you have generalized anxiety disorder (GAD),it is overwhelming and can dominate one's day with inappropriate and exaggerated worries and tension, even when there is nothing present to worry about. GAD is often accompanied by physical symptoms or sensations like a racing heart, breathlessness, nausea, can't sleep,chest pains and sweating for example.

Generalized anxiety disorder (GAD) involves anxiety and worry that is excessive and unrelenting. This high-level anxiety makes normal life difficult and relaxation impossible. If you have generalized anxiety disorder (GAD) you may worry about the same things that other people do: health issues, money, family problems, or difficulties at work. But you take these worries to a new level.

Normal worry vs. Generalized Anxiety Disorder (GAD)

"Normal" Worry:

- Your worrying doesn't get in the way of your daily activities and responsibilities.

- You're able to control your worrying.

- Your worries, while unpleasant, don't cause significant distress.

- Your worries are limited to a specific, small number of realistic concerns.

- Your bouts of worrying last for only a short time period.

Generalized Anxiety Disorder:

- Your worrying significantly disrupts your job, activities, or social life.

- Your worrying is uncontrollable.

- Your worries are extremely upsetting and stressful.

- You worry about all sorts of things, and tend to expect the worst.

- You've been worrying almost every day for at least six months.

GAD sufferers seem unable to eliminate or ignore their concerns, even though they usually realize that their anxiety is more intense than the situation warrants.

GAD sufferers may often feel light-headed or out of breath. They may also feel nauseated or have to go to the bathroom frequently. Generalized anxiety is constant and can cause anxiety/panic attacks during the day and night. Night time anxiety and panic attacks are especially disturbing and can often wake the sufferer from deep sleep feeling particularly frightened. Night time anxiety can be minimized with some practical steps which minimize the impact of such things as blood sugar level fluctuations during sleep.

Individuals with GAD seem unable to relax, and they may startle more easily than other people. They tend to have difficulty concentrating and often, they have trouble falling or staying asleep.

The sort of symptoms you can experience in generalized anxiety are generally milder than those experienced during an anxiety / panic attack. Although distressing, they are usually much less extreme but may include some of the symptoms outlined on our anxiety symptoms list.

Your stomach may churn, your heart races or beats slower or you may get palpitations you may also feel sweaty or clammy, dizzy or shaky and general unrest.

You may just feel as if you have the Flu with shaky or weak legs and clamminess. You might get disturbing thoughts or feel depressed, this is perfectly normal and will pass, it is purely a response to anxiety and must not be mistaken for depression. Remember, these thoughts and feelings are harmless but unpleasant none the less.

So let us look some levels of anxiety, such as: "Mum there's a great big spider in my bedroom" cries a child and the child gets anxious about the spider, a specific creature she imagines can hurt her. Anxiety always relates to hurt. This is a controllable anxiety, get rid of the spider equals getting rid of the anxiety, (carefully looking to see there are no other spiders lurking there). I would call this third-line-anxiety. A learned anxiety, overcome with reason and action. This anxiety is a learned one, where the growing person has learned that certain things can harm us and being little (in psyche if not body) there is a need for immediate help from others perceived as being stronger than the self. This is usually enough and can be used in a creative and useful manner. An example would be learning self-defence to ward off bullies at school. There is also the chronic worrier who perceives their worry as an anxiety state, and rightly so, we may say anxiety at a lesser level. I would call this second-line anxiety. Overcome with specific understanding of the self. This is also reasonably easy to overcome by getting to know who you really are. This is not as hard as you may think.

However there is also another anxiety, of varying degrees, one where it 'takes over' the emotions and won't go away. A terrifying emptiness of blackness filled with 'spiders' of our deep inner psyche's creation, not simply one in the bedroom that can be overcome by a bit of reason, muscle and mum's trusty broom. I would call this first-line anxiety. This is often very difficult to clear, not because it can't be cleared, but by the very nature of distrust on which all anxiety is founded. Anxiety is a very real but unconscious form of expression of the hurting child

Let's understand where anxiety comes from. This may be a bit heavy, but stay in there and you'll be supplied with information you need to give you a foundation of knowledge to underpin your new, productive actions. You see, all life must express itself in its own way. You are special part of that life and you have been since you were first united as one by the penetration of daddy's sperm into mummy's egg. The life then created was a NEW and powerful life. Very different to the first two living creatures (egg and sperm) and already eager to get stuck into life into mummy's womb where it forms into another new life (life meeting life again). Then you grow into a baby, and all the experiences of mother as she carries you are recorded in your little psyche. Then you are born. What an experience! Life (baby) meeting life (the world baby will live in).

Here is how it ideally works:

Essentially the growing baby wants mother's attentions at all times. Baby has already gone through the amazing life series of development of all the organs of the body and has broken from being dependent on the mother, has travelled the sometimes very scary birth canal and is eager to go on to new things. All this at an emotional level. Then baby goes through a series of transitional developments... firstly the need for acceptance. This is the prime essential and we all carry that through life to the grave. From acceptance baby moves to Being. A genuine, real little bundle of life that has father as a guarantee of strength and mother as an assurance of sustenance, both vital to the growing child. Then with mother and father there to sustain life as a right, baby moves to what can be called Wellbeing. From there on to Identification where baby identifies with mother to gain his personal identity and from father for his social identity. Further development for the well-parented child leads to Status as a person and then on to Production.

But what if there is a blockage somewhere along that ideal chain? What if mother does NOT come when called upon by baby's very real needs? What would happen if the normal cycle was BLOCKED, for whatever reason, at the first stage when mother was NOT there to supply her loving acceptance/ sustenance? The inner pain can be endured for a time, but then baby may feel that mother will never come and a panic starts to become a pathway within the developing psyche. This pathway has no name as it's an indefinable feeling,but it's REAL to baby and it starts to develop its own pathways within the psyche. There starts a despair, a hopelessness and the ability to wait for the second stage of BEING by mother's coming is weakened. Infantile death! Baby doesn't know this of course, and it's all pushed into the feeling unconscious to be drawn on at any time later in life when similar experiences are presented.

Here is the foundation of hopelessness, the substance for anxiety.

Let me give you an illustration:

A lady came to me with a terrible anxiety about money. She was single, attractive, intelligent. Money was a terrifying agent for her anxiety on a large scale, permeating all she did. However her older brother was a millionaire by age 21! Same parents, same parental love shared by both. Both accepted. What was the difference? It appeared that her father was a very successful businessman in a middle eastern country. His livelihood was taken away from him by an antagonistic government action. The very lives of the father, mother and son were seriously at risk so they fled to United States as immigrants with absolutely no money at all. The mother was devastated at all this and my client was conceived during this calamitous period. She wore her mother's grief (loss of livelihood) mother's desperation (lack of money) mother's shame (social scorn) and now mother's contribution to poverty (becoming pregnant) In US, the father resumed his business talents and again became wealthy, the son following dad's social example, but the girl only knew, among all the love shared in the family, the distress that associated with money. The son felt prosperity. That was HIS inner picture. The girl felt poverty-anxiety. THAT was her inner picture. She was able to step out of that state with the adult mind of CHOICE leaving behind the anxiety-feeling mind of the inner CHILD.

I will show you how you can do that also a little later.

At each stage of acceptance, being, wellbeing, identification, status, productivity, varying degrees of blocking can take place, leaving behind it's legacy of anxiety. Here is first line trauma. Bearing in mind that we are a complex of genetics, ego strength, ancestral and cellular memories, and family dynamics so it would be difficult indeed for a therapist to get to the bottom of it all.

So let's look at the basics of anxiety and as we go, if you feel (note the word feel) something stirring within you, take note of it, maybe write it down for your attention later on.

WE are a complex of many ingredients and all of these ingredients and circumstances join together to make the strength of our ego (who we believe ourselves to be). If a person has a strong ego they will act accordingly and if a weak ego (or none at all) they will act accordingly. Makes sense. But for the growing child the sense of power lies in the parenting, especially mother. So the little ego will react to the bad as well as the good in it's own strength. This is why you are unique. No one is the same as you and YOU have the inner answer to your inner anxiety.

Let's look at how it can operate:

Baby can, and does, suffer from mental anguish at the loss of BEING through the loss of mother's attentions, for whatever reason, (don't go blaming mum as it quite often, indeed almost always, someone else messing up her routine with baby) and this results in dread sometimes with rage attached and sometimes beyond rage with depressions as a result. Different babies, different ego-responses. What happens in this case is baby develops a feeling, nay, a TERROR of the death of the spirit. Baby (you and me) has been through.

The joys and energetic drives of LIFE! LIVING! Taking LIFE as LIFE itself with all of the energetic joys of getting 'stuck into life' in mother's endometrium. BLISS! And now being denied this feeling of LIFE! and an emerging sense of DEATH!

This once LIFE! Being somehow blocked and this blocking taking away that BEING so essential to baby's growth - a growth that cannot be stopped except by death. So: delusional fears of death (self or close persons). Being preoccupied with impending death. Doom. Disaster. Hell. So with this mind-set there is an entry into a panic state, intense apprehension, ANXIETY as a first-line anxiety.

A sense of despair 'takes over'. Various phobias are revealed that are peculiar to the experiences of the little baby/child. The ability to wait for 'the good' to come in the form of some accepting person is almost nil. The BEING is challenged by a feeling of separation that will never be satisfied. Therefore no sustenance. A hopelessness called anxiety. This feeling was once REAL and the feeling is still REAL to the inflicted person. However note that it is a FEELING and as such cannot harm you in any way. A harmless but intense feeling that simply 'takes over' as it did when baby was experiencing the FACT of non-BEING.

Anxiety at this level can lead to endless talking about their terrible plight. Feeling 'locked in' to endless bad relationships. Difficult to maintain true friendships, which fuels the anxiety. Shutting one's self off from help. Isolated by others. More anxiety.

Indeed there are so many paths that can be developed by the human psyche regarding anxiety we could wonder if there is ANY real, basic relief from it's malevolent attacks. Let's look at some of the expressions of this malady, remembering that if baby was caught in this terrifying experience the deepest REPRESSION covers this desperate infantile deprivation and can lead to the death of the spirit in dread. It is a basic TERROR.

Caught in a RAGE Catch 22: Not to rage as a baby/child is to die by isolation. To rage is to both destroy and be destroyed. Have you had that feeling? If so, this is your story. Anxiety releases infantile RAGE and dreams (nightmares?) of how to destroy the whole rotten anxiety mess. The mother delay becomes an intolerable bondage, forced into the unconscious, but wanting to DO something (fright, flight, fight) which translates into the physical counterparts of: tense head feelings, buzzing in the ears, head throbbing, nose often congested, vision blurred, tight chest, breathing irregular, heart beating fast, blood pressure up, loss of appetite, nervous dyspepsia, tight stomach, constipated, loss of weight, muscles tight and aching, pains in the neck and limbs, tremors, fidgeting, sweating, blotchy skin, menstruation preceded by tension, irritable, angry scenes, sexual libido often lost, and there are others. Survival is therefore by fight as mounting rage is an attempt to force a response. This is all an active aggressive response to this separation-anxiety.

Or we may go into the phobic areas of expression of our anxiety. there is only one anxiety, but many expressions of it: Agoraphobia, Claustrophobia, Fears of loss of consciousness, Hospitals, Separation, Being sick. etc.

Attention seeking can become a fixation with infantile dependent behaviour attached.

Any usual output is thereby lost, the energies being put to the inner world of the child within, unconsciously fighting from a platform of dire distress. Weakness or tiredness takes over. Not wanting to either start or finish a task.

Fatigue can become persistent and quite strong.

Loss of concentration. Can't read and remember what has been read. Little if any concern for others. Small problems become major events. Easily distracted. Antisocial. Not a party-goer. Only partial attempts to become party to activities.

Moods are gloomy with heavy, stony grief. Irritable, whining, always lamenting. All the good times are in the past. Memory selects past grievances. All is experienced as painful, even any happy times, and can't feel love for family or show or feel real grief.

Depersonalised and incapable of emotion. Unreality feelings include being 'outside myself' or 'not there'. Time drags. Colours look drab, or all is seen as grey.

The spirits are low, especially in the morning o after sleep. Negative people, Denies inner hostility. God is seen as distant, angry and rejecting.

As personalities before the trigger of anxiety in this level is activated, they could well have been well adjusted and outgoing. Others cold have been dependent and anxious with a record of hard work, anxiously, meticulously and over-conscientiously performed.

In this case they would have been denied mother, as a baby, but also had attentions not of their satisfaction, as a surrogate mother, or nurse. So you can see that there could be many variants and only YOU would have that, unconsciously, in your psyche.

Anxiety can cause physical and emotional symptoms. A specific situation or fear can cause some or all of these symptoms for a short time. When the situation passes, the symptoms usually go away.

Physical symptoms of anxiety include:

- Trembling, twitching, or shaking.

- Feeling of fullness in the throat or chest.

- Breathlessness or rapid heartbeat.

- Light-headedness or dizziness.

- Sweating or cold, clammy hands.

- Feeling jumpy.

- Muscle tension, aches, or soreness (myalgias).

- Extreme tiredness.

Sleep problems, such as the inability to fall asleep or stay asleep, early waking, or restlessness (not feeling rested when you wake up).

Anxiety affects the part of the brain that helps control how you communicate. This makes it harder to express yourself creatively or function effectively in relationships. Emotional symptoms of anxiety include:

- Restlessness, irritability, or feeling on edge or keyed up.

- Worrying too much.

- Fearing that something bad is going to happen; feeling doomed.

- Inability to concentrate; feeling like your mind goes blank.

Anxiety disorders

Anxiety disorders occur when people have both physical and emotional symptoms. Anxiety disorders interfere with how a person gets along with others and affect daily activities. Women are twice as likely as men to have problems with anxiety disorders. Examples of anxiety disorders include panic attacks, phobias, and generalized anxiety disorder.

Often the cause of anxiety disorders is not known. Many people with an anxiety disorder say they have felt nervous and anxious all their lives. This problem can occur at any age. Children who have at least one parent with the diagnosis of depression are more likely to have an anxiety disorder than other children.

Anxiety disorders often occur with other problems, such as:

Mental health problems, such as depression.

Substance use problems.

A physical problem, such as heart or lung disease. A complete medical examination may be needed before an anxiety disorder can be diagnosed.

Panic attacks

A panic attack is a sudden feeling of extreme anxiety or intense fear without a clear cause or when there is no danger. Panic attacks are common. They sometimes occur in otherwise healthy people. An attack starts suddenly and usually lasts from 5 to 20 minutes but may last even longer. You have the most anxiety about 10 minutes after the attack starts.

Symptoms include feelings of dying or losing control of yourself, rapid breathing (hyperventilation), numbness or tingling of the hands or lips, and a racing heart. You may feel dizzy, sweaty, or shaky. Other symptoms include trouble breathing, chest pain or tightness, and an irregular heartbeat. These symptoms come on suddenly and without warning.

Sometimes symptoms of a panic attack are so intense that the person fears he or she is having a heart attack. Many of the symptoms of a panic attack can occur with other illnesses, such as hyperthyroidism, coronary artery disease, or chronic obstructive pulmonary disease (COPD). A complete medical examination may be needed before an anxiety disorder can be diagnosed.

People who have repeated unexpected panic attacks and worry about the attacks are said to have a panic disorder.

Depression

While everyone feels sad from time to time, major depression is very different. Major depressive disorder or clinical depression causes you to experience feelings of sadness, loneliness, or a loss of interest in things you once enjoyed. When these feelings occur for more than two weeks, doctors may diagnose this as major depressive disorder. These symptoms are a sign that you need to seek professional help. Talk to your doctor if you have symptoms that may indicate depression.

Common symptoms of depression

Symptoms of depression can vary. They may manifest themselves differently from person to person. However, for most people, depression symptoms affect their ability to perform daily activities, interact with others, or go to work or go to school. If you suffer from depression you may often experience several of the following:

Sadness

The most common symptom of depression is a feeling of sadness or emptiness that lasts for more than two weeks. A person may describe this symptom as a feeling of "hopelessness." They may feel as if life will not get better and that this intense level of sadness will last forever. If this feeling lasts longer than two years it's known as dysthymia. This is a type of chronic depression in which a person's moods are consistently low.

Worthlessness

Continual feelings of worthlessness, guilt, or helplessness often accompany the condition. People tend to focus on personal shortcomings or past failures. They often blame themselves when their life isn't going the way they would like. Teenagers who experience depression commonly report feelings of worthlessness. They may report feeling misunderstood and start to avoid interactions with others.

Irritability

Depression may cause people to get easily frustrated or angered, even over small or insignificant matters. This often relates back to a person experiencing levels of tension and fatigue that makes it difficult to get through the day. Men and women may display irritability symptoms differently from each other. Women often report feeling angry at one moment, and then tearful at the next. Men may appear volatile or aggressive due to their depression. Traditional male roles in society may also mean that a man displays irritability for not being able to "get it together" and overcome depressive symptoms.

Fatigue

People with depression often experience lack of energy or feel tired all the time. Small tasks, like showering or getting out of bed, may seem to require more effort than one can muster. Fatigue can play a role in other symptoms associated with depression, such as withdrawal and apathy. You may feel overwhelmed at the mere thought of exertion or going outdoors.

Guilt

Depression is often the result of imbalanced chemicals in the brain. However, people experiencing depression may blame themselves for their symptoms instead. Statements such as "I can't do anything right" or "everything is my fault," become the norm for you.

Crying spells

People who have depression may find themselves crying frequently for no apparent reason. Crying spells can be a symptom of post-partum depression, which can occur in a woman after she's given birth.

Apathy

People with depression commonly lose interest or stop finding pleasure in activities that they once enjoyed, including sex.

Anxiety

Anxiety is a feeling of impending doom or danger, even when there isn't a justifiable reason. Depression can cause a person to feel anxious all the time. A person may say they are constantly tense, but there's no direct threat or identifiable source for this tension.

Restlessness

Agitation and restlessness, including pacing, an inability to sit still, or hand wringing, may occur with depression.

Lack of concentration

People with depression may have a difficult time remembering, maintaining focus, or making decisions. Fatigue, feelings of worthlessness, or feeling "numb" can turn decision-making into a talk that is difficult to accomplish. Friends or family members may discuss specific dates or events, but you may not remember just moments later due to concentrating lack of concentration. This inability to concentrate can lead to withdrawal in a depressed person.

Withdrawal

Many people with depression shut themselves off from the world. They may isolate themselves, not answer the phone, or refuse to go out with friends. You feel as if you're "numb," and that nothing will bring you joy.

Sleep problems

People's sleep habits are likely to change as a result of depression. They may not be able to fall asleep or stay asleep. They may wake up in the middle of the night and not go back to sleep at all. You may sleep for long periods and find that you don't want to get out of bed. These symptoms lead to fatigue that can exacerbate additional symptoms of depression, such as a lack of concentration.

Overeating or loss of appetite

Depression can often cause a lack of interest in food and weight loss. In other people, depression leads to overeating and weight gain. This is because a person may feel so frustrated or miserable that they turn to food as a means to escape their problems. However, overeating can lead to weight gain and cause you to exhibit low levels of energy. Not enough food can also cause you to also have low energy levels and feel weak.

Thoughts of suicide

Thinking or fantasizing about death is a serious sign that needs to be addressed right away. According to the Mayo Clinic, thoughts of suicide are symptoms common in older men. Loved ones may not initially notice this thinking and pass a person's depression symptoms off as age-related mental health changes. However, depression and especially suicidal thoughts are never normal emotions.

If you or a loved one is thinking of hurting themselves, seek immediate medical attention. At the emergency room, a doctor can help you get mental health care until these feelings subside.

Physical pain

Physical symptoms, such as body pain, headaches, cramps, and digestive problems also can occur. Younger children with depression commonly report physical pain symptoms. They may refuse to go to school or behave particularly clingy due to the worry about their aches and pains.

Phobias

Fear is a powerful and primitive human emotion. It alerts us to the presence of danger and it was critical in keeping our ancestors alive. Fear can be divided into two responses, biochemical and emotional. The biochemical response is universal, while the emotional response is highly individual.

Biochemical Reaction

Fear is a natural emotion and a survival mechanism. When we confront a perceived threat, our bodies respond in specific ways. Physical reactions to fear include sweating, increased heart rate, and high adrenaline levels that make us extremely alert. This physical response is also known as the "fight or flight" response, in which your body prepares itself to either enter combat or run away. This biochemical reaction is likely an evolutionary development. It's an automatic response that is crucial to our survival.

Emotional Response

The emotional response to fear is highly personalized. Because fear involves some of the same chemical responses in our brains that positive emotions like happiness and excitement do, feeling fear under certain circumstances can be seen as fun, like when you watch scary movies. Some people are adrenaline junkies, thriving on extreme sports and other fear-inducing thrill situations. Others have a negative reaction to the feeling of fear, avoiding fear-inducing situations at all costs. Although the physical reaction is the same, fear may be perceived as either positive or negative, depending on the person.

Causes of Fear

Fear is incredibly complex. Some fears may be a result of experiences or trauma, while others may actually represent a fear of something else entirely, such as a loss of control. Still other fears may occur because they cause physical symptoms, such as being afraid of heights because they make you feel dizzy and sick to your stomach, even if you're simply watching a video or looking at a picture and in no actual danger.

Scientists are trying to understand exactly what fear is and what causes it, but this is a supremely difficult undertaking in light of the differences between individuals in terms of what they fear and why. Not to mention that there is no agreement between scientists who study fear as to whether it's a sort of behavior that's only observable or something our brains are physically wired to do.

Fear is the basis of anxiety but it is not the only anxiety attack symptom. Treatment for your anxiety needs to address this fear first, however, because it is the trigger for anxiety.

You'll likely want to look at natural anxiety relief, rather than drug-induced treatments because at some point, you will want to be drug and anxiety free.

So let's talk about fear itself and how you can start treating yourself now.

Where does this fear come from? Why does it cause such distress and feed anxiety as well as it does? What can you do to stop it?

There have been many theories around fear and its role as an anxiety attack symptom. Treatment options are equally diverse. Some are scams while others have proven to be highly effective, regardless of how long you've suffered anxiety attack symptoms.

Having suffered years of anxiety, I underwent the related medical treatments. The medications merely kept me calm while I went through 3 years of therapy, but when it was over, the fear had eased but it hadn't gone away.

Then one day, I discovered the secret behind that fear. Today, I immediately recognize when the fear is stirring and can halt it immediately.

If you can identify your fear, you will be able to do the same.

First let me say that I disagree with the approach that you don't have to know what causes the fear, and that it's only important to know that the fear is simply an anxiety attack symptom. For me, it was imperative I figure out the fear so I could deal with it. And when I did, my anxiety attacks subsided.

I hope this secret I'm about to reveal helps you too.

The first thing to realize is that fear comes from your subconscious. Your body senses a danger or perceived danger and responds with fear. Fear causes your body to increase its adrenalin so it can fight to preserve itself. Today, however, much of the fear is internalized rather than from an outside source, like a threat from a wild animal.

It's true that over time our experiences have taught us to respond automatically with this fear instinct. When it gets out of control, as when it becomes an anxiety attack symptom, treatment must be sought.

Let's start with your thoughts, because that's where the fear hides.

If you listen closely when you feel an anxiety attack coming on, you'll get a sense that something just happened to scare you. You know that feeling you get when you go into a room and forget why you went there? Fear as an anxiety attack symptom feels much the same. Inside, you feel uncomfortable. It's a feeling that's difficult to pinpoint or explain. You just sense something isn't right.

I firmly believe that the fear builds because now you are afraid of the fear itself, just as popular programs say. You don't understand where it's coming from and this makes you believe something terrible is going to happen.

This 'something' that caused your unease might be a memory your subconscious has suppressed. For instance, let's say you had a bad experience years ago when you had to go to a meeting. Perhaps you were asked to make a brief presentation, but you hadn't been notified in advance and weren't prepared. You immediately felt fear. This is a natural response to such an unexpected event.

One day, you're going about your business and something reminds you of that nerve-wracking event and the fear you felt resurfaces, but you try to suppress it. This time, you can't pinpoint what's causing it. You have now set up an automatic fear response that will continue until you find a way to deal with it.

When you sense your fear, stop and think about what just happened. What were you thinking about? Are you anxious about something in your life? Often, we have day-to-day 'fears' about things. We worry over finances, health, career, family, retirement, an expected move.

Perhaps you're having trouble meeting your bill payments. Maybe you have to do something you don't particularly feel comfortable doing. Maybe you are going somewhere and you're nervous about the trip for some reason.

Take time to sit down and write out the things that make you nervous, fearful, stressed. Cover your personal traits - for example, if you're a homebody and would rather keep to yourself than attend a meeting or go to a social gathering. Include any traumatic events in your life, going right back to the beginning.

Have you lost loved ones recently? Are you concerned about your life without them? This is particularly tough and might require help from bereavement support groups. Be sure to take advantage of them, because they can provide tremendous help in moving on.

How's your self esteem? Do you feel good about yourself? You might have to really think critically to discover just who you are and why you are the way you are. Write it all down and see if you can identify the things that trigger your fear.

See, my occasional lack of confidence will make me uncertain when I have to take on certain tasks. I'm not sure I can handle them, or that I'll do a good job. Now I know this is a threat to my self-esteem. If I fail, it shows I'm not as good as I'd like to believe.

Failing only reveals my weaknesses and limitations. Because self-esteem is necessary to live comfortably, fear results from this 'threat'. And yes, that fear can soon turn into an anxiety attack if I don't take those few minutes to catch those fleeting thoughts that caused it.

First you need to work on fear - your anxiety attack symptom. Treatment begins at home, as they say.

I've discovered that just the act of halting my fear to stop and think about what just went through my mind was enough to bring the anxiety attack to an end in a matter of seconds. As soon as that happened, I was able to think about the fear and focus on solutions, rather than on the fear itself.

Have you ever wakened from a dream and immediately forgotten what it was, but it felt important to remember? You struggled to recall it so you could find out what happened or resolve the problem. Think of your fear as a dream - something in your subconscious that you can resolve.

Remember too, that most fear is fear of the unknown. Identify your fear and it no longer has control over you. You have control over it, and your anxiety.

I wish I could better explain this process, but the best I can advise is to watch your thinking. Listen to yourself, to your thoughts. Understand who you are. Take the time to sit down and truly evaluate what makes you tick.

Be honest in your personal evaluation. It can be difficult, but it's absolutely necessary for this to work. Often we don't recognize how we think, especially when we're talking about the subconscious. For instance, you might believe you have good self esteem and that you're upbeat and forward-thinking. You can be almost certain this is not entirely true in all situations. Ask your partner or family member for feedback if necessary. Sometimes an outside source can pick up things you'll miss.

Acclimation

Repeated exposure to similar situations leads to familiarity. This greatly reduces both the fear response and the resulting elation, leading adrenaline junkies to seek out ever new and bigger thrills. It also forms the basis of some phobia treatments, which depend on slowly minimizing the fear response by making it feel familiar.

Psychology of Phobias

One aspect of anxiety disorders can be a tendency to develop a fear of fear. Where most people tend to experience fear only during a situation that is perceived as scary or threatening, those who suffer from anxiety disorders may become afraid that they will experience a fear response. They perceive their fear responses as negative and go out of their way to avoid those responses.

A phobia is a twisting of the normal fear response. The fear is directed toward an object or situation that does not present a real danger. Though you recognize that the fear is unreasonable, you can't help the reaction. Over time, the fear tends to worsen as the fear of fear response takes hold.

CHAPTER TWO

BECOME FAMILIAR WITH THE BASIC PRINCIPLES OF CBT AND UNDERSTAND HOW IT WORKS

The History Of Cognitive Behavioral Therapy

Cognitive behavioral therapy is an approach used by psychotherapists to influence a patient's behaviors and emotions. The key to the approach is in its procedure which must be systematic. It has been used successfully to treat a variety of disorders including eating disorders, substance abuse, anxiety and personality disorders. It can be used in individual or group therapy sessions and the approach can also be geared towards self help therapy.

Cognitive behavioral therapy is a combination of traditional behavioral therapy and cognitive therapy. They are combined into a treatment that is focused on symptom removal. The effectiveness of the treatment can clearly be judged based on its results. The more it is used, the more it has become recommended. It is now used as the number one treatment technique for post traumatic stress disorder, obsessive compulsive disorder, depression and bulimia.

Cognitive behavioral therapy first began to be used between 1960 and 1970. It was a gradual process of merging behavioral therapy techniques and cognitive therapy techniques. Behavioral therapy had been around since the 1920's, but cognitive therapy was not introduced until the 1960's. Almost immediately the benefits of combining it with behavioral therapy techniques were realized. Ivan Pavlov, with his dogs who salivated at the ringing of the dinner bell, was among the most famous of the behavioral research pioneers. Other leaders in the field included John Watson and Clark Hull.

Instead of focusing on analyzing the problem like Freud and the psychoanalysts, cognitive behavioral therapy focused on eliminating the symptoms. The idea being that if you eliminate the symptoms, you have eliminated the problem. This more direct approach was seen as more effective at getting to the problem at hand and helping patients to make progress more quickly.

As a more radical aggressive treatment, behavioral techniques dealt better with more radical problems. The more obvious and clear cut the symptoms were, the easier it was to target them and devise treatments to eliminate them. Behavioral therapy was not as successful initially with more ambiguous problems such as depression. This realm was better served with cognitive therapy techniques.

In many academic settings, the two therapy techniques were used side by side to compare and contrast the results. It was not long before the advantages of combining the two techniques became clear as a way of taking advantage of the strengths of each. David Barlow's work on panic disorder treatments provided the first concrete example of the success of the combined strategies.

Cognitive behavioral therapy is difficult to define in a succinct definition because it covers such a broad range of topics and techniques. It is really an umbrella definition for individual treatments that are specifically tailored to the problems of a specific patient. So the problem dictates the specifics of the treatment, but there are some common themes and techniques. These include having the patient keep a diary of important events and record the feelings and behaviors they had in association with each event. This tool is then used as a basis to analyze and test the patient's ability to evaluate the situation and develop an appropriate emotional response. Negative emotions and behaviors are identified as well as the evaluations and beliefs that lead to them. An effort is then made to counter these beliefs and evaluations to show that the resulting behaviors are wrong. Negative behaviors are eliminated and the patient is taught a better way to view and react to the situation.

Part of the therapy also includes teaching the patient ways to distract themselves or change their focus from something that is upsetting or a situation that is generating negative behavior. They learn to focus on something else instead of the negative stimulus, thus eliminating the negative behavior that it would lead to. The problem is essentially nipped in the bud. For serious psychological disorders like bipolar disorder or schizophrenia, mood stabilizing medications are often prescribed to use in conjunction with these techniques. The medications give the patient enough of a calming effect to give them the opportunity to examine the situation and make the healthy choice whereas before they could not even pause for rational thought.

Cognitive behavioral therapy has been proven effective for a variety of problems, but it is still a process, not a miracle cure. It takes time to teach patients to understand situations and identify the triggers of their negative behaviors. Once this step is mastered, it still takes a lot of effort to overcome their first instincts and instead stop and make the right choices. First they learn what they should do, and then they must practice until they can do it.

The Basics of Cognitive Behavioral Therapy

The underlying concept behind CBT is that our thoughts and feelings play a fundamental role in our behavior. For example, a person who spends a lot of time thinking about plane crashes, runway accidents and other air disasters may find themselves avoiding air travel.

The goal of cognitive behavior therapy is to teach patients that while they cannot control every aspect of the world around them, they can take control of how they interpret and deal with things in their environment.

Cognitive behavior therapy has become increasingly popular in recent years with both mental health consumers and treatment professionals. Because CBT is usually a short-term treatment option, it is often more affordable than some other types of therapy. CBT is also empirically supported and has been shown to effectively help patients overcome a wide variety of maladaptive behaviors.

Cognitive behavioral therapy is a psychotherapeutic approach that aims to teach a person new skills on how to solve problems concerning dysfunctional emotions, behaviors, and cognitions through a goal-oriented, systematic approach. This title is used in many ways to differentiate behavioral therapy, cognitive therapy, and therapy that is based on both behavioral and cognitive therapies. There is empirical evidence that shows that cognitive behavioral therapy is quite effective in treating several conditions, including personality, anxiety, mood, eating, substance abuse, and psychotic disorders. Treatment is often manualized, as specific psychological orders are treated with specific technique-driven brief, direct, and time-limited treatments.

Cognitive behavioral therapy can be used both with individuals and in groups. The techniques are often adapted for self-help sessions as well. It is up to the individual clinician or researcher on whether he/she is more cognitive oriented, more behavioral oriented, or a combination of both, as all three methods are used today. Cognitive behavioral therapy was born out of a combination of behavioral therapy and cognitive therapy. These two therapies have many differences, but found common ground on focusing on the "here and now" and on alleviating symptoms.

Evaluating cognitive behavioral therapy has led to many believing that it is more effective over psychodynamic treatments and other methods. The United Kingdom advocates the use of cognitive behavioral therapy over other methods for many mental health difficulties, including post-traumatic stress disorder, obsessive-compulsive disorder, bulimia nervosa, clinical depression, and the neurological condition chronic fatigue syndrome/myalgic encephalomyelitis. The precursors of cognitive behavioral therapy base their roots in various ancient philosophical traditions, especially Stoicism. The modern roots of CBT can be traced to the development of behavioral therapy in the 1920s, the development of cognitive therapy in the 1960s, and the subsequent merging of the two therapies. The first behavioral therapeutic approaches were published in 1924 by Mary Cover Jones, whose work dealt with the unlearning of fears in children.

The early behavioral approaches worked well with many of the neurotic disorders, but not so much with depression. Behavioral therapy was also losing in popularity due to the "cognitive revolution." This eventually led to cognitive therapy being founded by Aaron T. Beck in the 1960s. The first form of cognitive behavioral therapy was developed by Arnold A. Lazarus during the time period of the late 1950s through the 1970s. During the 1980s and 1990s, cognitive and behavioral therapies were combined by work done by David M. Clark in the United Kingdom and David H. Barlow in the United States. Cognitive behavioral therapy includes the following systems: cognitive therapy, rational emotive behavior therapy, and multimodal therapy. One of the greatest challenges is defining exactly what a cognitive-behavioral therapy is. The particular therapeutic techniques vary within the different approaches of

CBT depending upon what kind of problem issues are being dealt with, but the techniques usually center around the following:

- Keeping a diary of significant events and associated feelings, thoughts, and behaviors.

- Questioning and testing cognitions, evaluations, assumptions, and beliefs that might be unrealistic and unhelpful.

- Gradually facing activities that may have been avoided.

- Trying out new ways of behaving and reacting.

In addition, distraction techniques, mindfulness, and relaxation are also commonly used in cognitive behavioral therapy. Mood-stabilizing medications are also often combined with therapies to treat conditions like bipolar disorder. The NICE guidelines within the British NHS recognize cognitive behavioral therapy's application in treating schizophrenia in combination with medication and therapy. Cognitive behavioral therapy usually takes time for patients to effectively implement it into their lives. It usually takes concentrated effort for them to replace a dysfunctional cognitive-affective-behavioral process or habit with a more reasonable and adaptive one, even when they recognize when and where their mental processes go awry. Cognitive behavioral therapy is applied to many different situations, including the following conditions:

Anxiety disorders (obsessive-compulsive disorder, social phobia or social anxiety, generalized anxiety disorder)

Mood disorders (clinical depression, major depressive disorder, psychiatric symptoms)

Insomnia (including being more effective than the drug Zopiclone)

Severe mental disorders (schizophrenia, bipolar disorder, severe depression)

Children and adolescents (major depressive disorder, anxiety disorders, trauma and posttraumatic stress disorder symptoms)

Stuttering (to help them overcome anxiety, avoidance behaviors, and negative thoughts about themselves)

Cognitive behavioral therapy involves teaching a person new skills to overcome dysfunctional emotions, behaviors, and cognitions through a goal-oriented, systematic approach. There is empirical evidence showing that cognitive behavioral therapy is effective in treating many conditions, including obsessive-compulsive disorder, generalized anxiety disorder, major depressive disorder, schizophrenia, anxiety, and negative thoughts about oneself). With the vast amount of success shown by the use of this therapy, it is one of the most important tools that researchers and therapists have to treat mental disorders today.

Automatic Negative Thoughts

One of the main focuses of cognitive-behavioral therapy is on changing the automatic negative thoughts that can contribute to and exacerbate emotional difficulties, depression, and anxiety. These negative thoughts spring forward spontaneously, are accepted as true, and tend to negatively influence the individual's mood.

Through the CBT process, patients examine these thoughts and are encouraged to look at evidence from reality that either supports or refutes these thoughts. By doing this, people are able to take a more objective and realistic look at the thoughts that contribute to their feelings of anxiety and depression. By becoming aware of the negative and often unrealistic thoughts that dampen their feelings and moods, people are able to start engaging in healthier thinking patterns.

Types of Cognitive Behavior Therapy

According to the British Association of Behavioural and Cognitive Psychotherapies, "Cognitive and behavioral psychotherapies are a range of therapies based on concepts and principles derived from psychological models of human emotion and behavior. They include a wide range of treatment approaches for emotional disorders, along a continuum from structured individual psychotherapy to self-help material."

There are a number of specific types of therapeutic approaches that involve CBT that are regularly used by mental health professionals. Examples of these include:

Rational Emotive Behavior Therapy (REBT): This type of CBT is centered on identifying and altering irrational beliefs. The process of REBT involves identifying the underlying irrational beliefs, actively challenging these beliefs, and finally learning to recognize and change these thought patterns.

Cognitive Therapy: This form of therapy is centered on identifying and changing inaccurate or distorted thinking patterns, emotional responses, and behaviors.

Multimodal Therapy: This form of CBT suggests that psychological issues must be treated by addressing seven different but interconnected modalities, which are behavior, affect, sensation, imagery, cognition, interpersonal factors and drug/biological considerations.

Dialectical Behavior Therapy: This type of cognitive-behavioral therapy addresses thinking patterns and behaviors and incorporates strategies such as emotional regulation and mindfulness.

While each type of cognitive-behavioral therapy offers its own unique approach, each centers on addressing the underlying thought patterns that contribute to psychological distress.

The Components of Cognitive Behavior Therapy

People often experience thoughts or feelings that reinforce or compound faulty beliefs. Such beliefs can result in problematic behaviors that can affect numerous life areas, including family, romantic relationships, work, and academics.

For example, a person suffering from low self-esteem might experience negative thoughts about his or her own abilities or appearance. As a result of these negative thinking patterns, the individual might start avoiding social situations or pass up opportunities for advancement at work or at school.

In order to combat these destructive thoughts and behaviors, a cognitive-behavioral therapist begins by helping the client to identify the problematic beliefs. This stage, known as functional analysis, is important for learning how thoughts, feelings, and situations can contribute to maladaptive behaviors. The process can be difficult, especially for patients who struggle with introspection, but it can ultimately lead to self-discovery and insights that are an essential part of the treatment process.

The second part of cognitive behavior therapy focuses on the actual behaviors that are contributing to the problem. The client begins to learn and practice new skills that can then be put in to use in real-world situations. For example, a person suffering from drug addiction might start practicing new coping skills and rehearsing ways to avoid or deal with social situations that could potentially trigger a relapse.

In most cases, CBT is a gradual process that helps a person take incremental steps towards a behavior change. Someone suffering from social anxiety might start by simply imagining himself in an anxiety-provoking social situation.

Next, the client might start practicing conversations with friends, family, and acquaintances. By progressively working toward a larger goal, the process seems less daunting and the goals easier to achieve.

The Process of Cognitive Behavior Therapy

During the process of CBT, the therapist tends to take a very active role.

CBT is highly goal-oriented and focused, and the client and therapist work together as collaborators toward the mutually established goals.

The therapist will typically explain the process in detail and the client will often be given homework to complete between sessions.

Cognitive-behavior therapy can be effectively used as a short-term treatment centered on helping the client deal with a very specific problem.

Uses of Cognitive Behavior Therapy

Cognitive behavior therapy has been used to treat people suffering from a wide range of disorders, including:

1. Anxiety
2. Phobias
3. Depression
4. Addictions
5. Eating disorders
6. Panic attacks
7. Anger

CBT is one of the most researched types of therapy, in part because treatment is focused on highly specific goals and results can be measured relatively easily.

Compared to psychoanalytic types of psychotherapy which encourage a more open-ended self-exploration, cognitive behavior therapy is often best-suited for clients who are more comfortable with a structured and focused approach in which the therapist often takes an instructional role. However, for CBT to be effective, the individual must be ready and willing to spend time and effort analyzing his or her thoughts and feelings. Such self-analysis and homework can be difficult, but it is a great way to learn more about how internal states impact outward behavior.

Cognitive behavior therapy is also well-suited for people looking for a short-term treatment option for certain types of emotional distress that does not necessarily involve psychotropic medication. One of the greatest benefits of cognitive-behavior therapy is that it helps clients develop coping skills that can be useful both now and in the future.

Although therapy must be tailored to the individual, there are, nevertheless, certain principles that underlie cognitive behavior therapy for all patients. I will use a depressed patient, "Amy," to illustrate these central tenets and to demonstrate how to use cognitive theory to understand patients' difficulties and how to use this understanding to plan treatment and conduct therapy sessions.

Amy was an 18-year-old single female when she sought treatment with me during her second semester of college. She had been feeling quite depressed and anxious for the previous 4 months and was having difficulty with her daily activities. She met criteria for a major depressive episode of moderate severity according to DSM-IV-TR (the Diagnostic and Statistical Manual of Mental Disorders, Fourth Edition, Text Revision; American Psychiatric Association, 2000). The basic principles of cognitive behavior therapy are as follows:

Principle No. 1: Cognitive behavior therapy is based on an ever-evolving formulation of patients' problems and an individual conceptualization of each patient in cognitive terms. I consider Amy's difficulties in three time frames. From the beginning, I identify her current thinking that contributes to her feelings of sadness ("I'm a failure, I can't do anything right, I'll never be happy"), and her problematic behaviors (isolating herself, spending a great deal of unproductive time in her room, avoiding asking for help). These problematic behaviors both flow from and in turn reinforce Amy's dysfunctional thinking.

Second, I identify precipitating factors that influenced Amy's perceptions at the onset of her depression(e.g., being away from home for the first time and struggling in her studies contributed to her belief that she was incompetent).

Third, I hypothesize about key developmental events and her enduring patterns of interpreting these events that may have predisposed her to depression (e.g., Amy has had a lifelong tendency to attribute personal strengths and achievement to luck, but views her weaknesses as a reflection of her "true" self).

I base my conceptualization of Amy on the cognitive formulation of depression and on the data Amy provides at the evaluation session. I continue to refine this conceptualization at each session as I obtain more data. At strategic points, I share the conceptualization with Amy to ensure that it "rings true" to her. Moreover, throughout therapy I help Amy view her experience through the cognitive model. She learns, for example, to identify the thoughts associated with her distressing affect and to evaluate and formulate more adaptive responses to her thinking. Doing so improves how she feels and often leads to her behaving in a more functional way.

Principle No. 2: Cognitive behavior therapy requires a sound therapeutic alliance. Amy, like many patients with uncomplicated depression and anxiety disorders, has little difficulty trusting and working with me. I strive to demonstrate all the basic ingredients necessary in a counseling situation: warmth, empathy, caring, genuine regard, and competence. I show my regard for Amy by making empathic statements, listening closely and carefully, and accurately summarizing her thoughts and feelings. I point out her small and larger successes and maintain a realistically optimistic and upbeat outlook. I also ask Amy for feedback at the end of each session to ensure that she feels understood and positive about the session.

Principle No. 3: Cognitive behavior therapy emphasizes collaboration and active participation. I encourage Amy to view therapy as teamwork; together we decide what to work on each session, how often we should meet, and what Amy can do between sessions for therapy homework. At first, I am more active in suggesting a direction for therapy sessions and in summarizing what we've discussed during a session. As Amy becomes less depressed and more socialized into treatment, I encourage her to become increasingly active in the therapy session: deciding which problems to talk about, identifying the distortions in her thinking, summarizing important points, and devising homework assignments.

Principle No. 4: Cognitive behavior therapy is goal oriented and problem focused. I ask Amy in our first session to enumerate her problems and set specific goals so both she and I have a shared understanding of what she is working toward. For example, Amy mentions in the evaluation session that she feels isolated. With my guidance, Amy states a goal in behavioral terms: to initiate new friendships and spend more time with current friends. Later, when discussing how to improve her day-to-day routine, I help her evaluate and respond to thoughts that interfere with her goal, such as: My friends won't want to hang out with me. I'm too tired to go out with them. First, I help Amy evaluate the validity of her thoughtsthrough an examination of the evidence. Then Amy is willing to test the thoughts more directly through behavioral experiments in which she initiates plans with friends. Once she recognizes and corrects the distortion in her thinking, Amy is able to benefit from straightforward problem solving to decrease her isolation.

Principle No. 5: Cognitive behavior therapy initially emphasizes the present. The treatment of most patients involves a strong focus on current problems and on specific situations that are distressing to them. Amy begins to feel better once she is able to respond to her negative thinking and take steps to improve her life. Therapy starts with an examination of here-and-now problems, regardless of diagnosis. Attention shifts to the past in two circumstances: One, when patients express a strong preference to do so, and a failure to do so could endanger the therapeutic alliance. Two, when patients get "stuck" in their dysfunctional thinking, and an understanding of the childhood roots of their beliefs can potentially help them modify their rigid ideas. ("Well, no wonder you still believe you're incompetent. Can you see how almost any child—who had the same experiences as you—would grow up believing she was incompetent, and yet it might not be true, or certainly not completely true?")

For example, I briefly turn to the past midway through treatment to help Amy identify a set of beliefs she learned as a child: "If I achieve highly, it means I'm worthwhile," and "If I don't achieve highly, it means I'm a failure." I help her evaluate the validity of these beliefs both in the past and present. Doing so leads Amy, in part, to develop more functional and more reasonable beliefs. If Amy had had a personality disorder, I would have spent proportionally more time discussing her developmental history and childhood origin of beliefs and coping behaviors.

Principle No. 6: Cognitive behavior therapy is educative, aims to teach the patient to be her own therapist, and emphasizes relapse prevention. In our first session I educate Amy about the nature and course of her disorder, about the process of cognitive behavior therapy, and about the cognitive model (i.e., how her thoughts influence her emotions and behavior). I not only help Amy set goals, identify and evaluate thoughts and beliefs, and plan behavioral change, but I also teach her how to do so. At each session I ensure that Amy takes home therapy notes—important ideas she has learned—so she can benefit from her new understanding in the ensuing weeks and after treatment ends.

Principle No. 7: Cognitive behavior therapy aims to be time limited. Many straightforward patients with depression and anxiety disorders are treated for six to 14 sessions. Therapists' goals are to provide symptom relief, facilitate a remission of the disorder, help patients resolve their most pressing problems, and teach them skills to avoid relapse. Amy initially has weekly therapy sessions. (Had her depression been more severe or had she been suicidal, I may have arranged more frequent sessions.) After 2 months, we collaboratively decide to experiment with biweekly sessions, then with monthly sessions. Even after termination, we plan periodic "booster" sessions every 3 months for a year. Not all patients make enough progress in just a few months, however. Some patients require 1 or 2 years of therapy (or possibly longer) to modify very rigid dysfunctional beliefs and patterns of behavior that contribute to their chronic distress. Other patients with severe mental illness may need periodic treatment for a very long time to maintain stabilization.

Principle No. 8: Cognitive behavior therapy sessions are structured. No matter what the diagnosis or stage of treatment, following a certain structure in each session maximizes efficiency and effectiveness. This structure includes an introductory part (doing a mood check, briefly reviewing the week, collaboratively setting an agenda for the session), a middle part (reviewing homework, discussing problems on the agenda, setting new homework, summarizing), and a final part (eliciting feedback). Following this format makes the process of therapy more understandable to patients and increases the likelihood that they will be able to do self-therapy after termination.

Principle No. 9: Cognitive behavior therapy teaches patients to identify, evaluate, and respond to their dysfunctional thoughts and beliefs. Patients can have many dozens or even hundreds of automatic thoughts a day that affect their mood, behavior, or physiology (the last is especially pertinent to anxiety). Therapists help patients identify key cognitions and adopt more realistic, adaptive perspectives, which leads patients to feel better emotionally, behave more functionally, or decrease their physiological arousal. They do so through the process of guided discovery, using questioning (often labeled or mislabeled as "Socratic questioning") to evaluate their thinking (rather than persuasion, debate, or lecturing). Therapists also create experiences, called behavioral experiments, for patients to directly test their thinking (e.g., "If I even look at a picture of a spider, I'll get so anxious I won't be able to think"). In these ways, therapists engage in collaborative empiricism. Therapists do not generally know in advance to what degree a patient's automatic thought is valid or invalid, but together they test the patient's thinking to develop more helpful and accurate responses.

When Amy was quite depressed, she had many automatic thoughts throughout the day, some of which she spontaneously reported and others that I elicited (by asking her what was going through her mind when she felt upset or acted in a dysfunctional manner). We often uncovered important automatic thoughts as we were discussing one of Amy's specific problems, and together we investigated their validity and utility. I asked her to summarize her new viewpoints, and we recorded them in writing so that she could read these adaptive responses throughout the week to prepare her for these or similar automatic thoughts. I did not encourage her to uncritically adopt a more positive viewpoint, challenge the validity of her automatic thoughts, or try to convince her that her thinking was unrealistically pessimistic. Instead we engaged in a collaborative exploration of the evidence.

Principle No. 10: Cognitive behavior therapy uses a variety of techniques to change thinking, mood, and behavior. Although cognitive strategies such as Socratic questioning and guided discovery are central to cognitive behavior therapy, behavioral and problem-solving techniques are essential, as are techniques from other orientations that are implemented within a cognitive framework. For example, I used Gestalt-inspired techniques to help

Amy understand how experiences with her family contributed to the development of her belief that she was incompetent. I use psychodynamically inspired techniques with some Axis II patients who apply their distorted ideas about people to the therapeutic relationship. The types of techniques you select will be influenced by your conceptualization of the patient, the problem you are discussing, and your objectives for the session.

These basic principles apply to all patients. Therapy does, however, vary considerably according to individual patients, the nature of their difficulties, and their stage of life, as well as their developmental and intellectual level, gender, and cultural background. Treatment also varies depending on patients' goals, their ability to form a strong therapeutic bond, their motivation to change, their previous experience with therapy, and their preferences for treatment, among other factors. The emphasis in treatment also depends on the patient's particular disorder(s). Cognitive behavior therapy for panic disorder involves testing the patient's catastrophic misinterpretations (usually life- or sanity-threatening erroneous predictions) of bodily or mental sensations. Anorexia requires a modification of beliefs about personal worth and control. Substance abuse treatment focuses on negative beliefs about the self and facilitating or permission-granting beliefs about substance use.

How does it work?

Cognitive behavioural therapy (CBT) can help you make sense of overwhelming problems by breaking them down into smaller parts.

Some forms of psychotherapy focus on looking into the past to gain an understanding of current feelings. In contrast, CBT focuses on present thoughts and beliefs.

CBT can help people with many problems where thoughts and beliefs are critical. It emphasizes the need to identify, challenge, and change how a person views a situation.

According to CBT, people's pattern of thinking is like wearing a pair of glasses that makes us see the world in a specific way. CBT makes us more aware of how these thought patterns create our reality and determine how we behave.

In CBT, problems are broken down into five main areas:

- situations
- thoughts
- emotions
- physical feelings
- actions

CBT is based on the concept of these five areas being interconnected and affecting each other. For example, your thoughts about a certain situation can often affect how you feel both physically and emotionally, as well as how you act in response.

How CBT is different

CBT differs from many other psychotherapies because it's:

Pragmatic – it helps identify specific problems and tries to solve them

Highly structured – rather than talking freely about your life, you and your therapist discuss specific problems and set goals for you to achieve

Focused on current problems – it's mainly concerned with how you think and act now rather than attempting to resolve past issues

Collaborative – your therapist won't tell you what to do; they'll work with you to find solutions to your current difficulties

Stopping negative thought cycles

There are helpful and unhelpful ways of reacting to a situation, often determined by how you think about them.

For example, if your marriage has ended in divorce, you might think you've failed and that you're not capable of having another meaningful relationship.

This could lead to you feeling hopeless, lonely, depressed and tired, so you stop going out and meeting new people. You become trapped in a negative cycle, sitting at home alone and feeling bad about yourself.

But rather than accepting this way of thinking you could accept that many marriages end, learn from your mistakes and move on, and feel optimistic about the future.

This optimism could result in you becoming more socially active and you may start evening classes and develop a new circle of friends.

This is a simplified example, but it illustrates how certain thoughts, feelings, physical sensations and actions can trap you in a negative cycle and even create new situations that make you feel worse about yourself.

CBT aims to stop negative cycles such as these by breaking down things that make you feel bad, anxious or scared. By making your problems more manageable, CBT can help you change your negative thought patterns and improve the way you feel.

CBT can help you get to a point where you can achieve this on your own and tackle problems without the help of a therapist.

Exposure therapy

Exposure therapy is a form of CBT particularly useful for people with phobias or obsessive compulsive disorder (OCD).

In such cases, talking about the situation isn't as helpful and you may need to learn to face your fears in a methodical and structured way through exposure therapy.

Exposure therapy involves starting with items and situations that cause anxiety, but anxiety that you feel able to tolerate. You need to stay in this situation for one to two hours or until the anxiety reduces for a prolonged period by a half.

Your therapist will ask you to repeat this exposure exercise three times a day. After the first few times, you'll find your anxiety doesn't climb as high and doesn't last as long.

You'll then be ready to move to a more difficult situation. This process should be continued until you have tackled all the items and situations you want to conquer.

Exposure therapy may involve spending six to 15

hours with the therapist, or can be carried out using self-help books or computer programs. You'll need to regularly practice the exercises as prescribed to overcome your problems.

CBT sessions

CBT can be carried out with a therapist in one-to-one sessions or in groups with other people in a similar situation to you.

If you have CBT on an individual basis, you'll usually meet with a CBT therapist for between five and 20 weekly or fortnightly sessions, with each session lasting 30-60 minutes.

Exposure therapy sessions usually last longer to ensure your anxiety reduces during the session. The therapy may take place:

In a clinic

Outside – if you have specific fears there

In your own home – particularly if you have agoraphobia or OCD involving a specific fear of items at home

Your CBT therapist can be any healthcare professional who has been specially trained in CBT, such as a psychiatrist, psychologist, mental health nurse or GP.

First sessions

The first few sessions will be spent making sure CBT is the right therapy for you, and that you're comfortable with the process. The therapist will ask questions about your life and background.

If you're anxious or depressed, the therapist will ask whether it interferes with your family, work and social life. They'll also ask about events that may be related to your problems, treatments you've had, and what you would like to achieve through therapy.

If CBT seems appropriate, the therapist will let you know what to expect from a course of treatment. If it's not appropriate, or you don't feel comfortable with it, they can recommend alternative treatments.

Further sessions

After the initial assessment period, you'll start working with your therapist to break down problems into their separate parts. To help with this, your therapist may ask you to keep a diary or write down your thought and behaviour patterns.

You and your therapist will analyse your thoughts, feelings and behaviours to work out if they're unrealistic or unhelpful and to determine the effect they have on each other and on you. Your therapist will be able to help you work out how to change unhelpful thoughts and behaviours.

After working out what you can change, your therapist will ask you to practise these changes in your daily life.

This may involve:

Questioning upsetting thoughts and replacing them with more helpful ones

Recognizing when you're going to do something that will make you feel worse and instead doing something more helpful

You may be asked to do some "homework" between sessions to help with this process.

At each session, you'll discuss with your therapist how you've got on with putting the changes into practice and what it felt like. Your therapist will be able to make other suggestions to help you.

Confronting fears and anxieties can be very difficult. Your therapist won't ask you to do things you don't want to do and will only work at a pace you're comfortable with. During your sessions, your therapist will check you're comfortable with the progress you're making.

One of the biggest benefits of CBT is that after your course has finished, you can continue to apply the principles learned to your daily life. This should make it less likely that your symptoms will return.

CHAPTER THREE

HOW TO DEFINE A SPECIFIC GOAL TO WORK TOWARD OVER THE COURSE OF 6 WEEKS

Because of CBT's success, some of its tenants have become well-known. You might already have heard, for example, that thoughts affect feelings, or that behavioural change can affect negative thoughts.

But what other Cognitive Behavioral Therapy techniques and tools are there, and how do they help you? How are these CBT techniques used in a session?

AGENDA SETTING IN CBT

This is a collaborative process between your therapist and you to determine how to best make use of each session. At the start of each appointment both you and your therapist suggest items you'd like to discuss. A decision is then made on the order the points will be discussed, and how much time each one needs.

The point of agenda setting is to make sure that the session is well spent, and that the hour isn't lost to something that isn't of itself productive, like simply rehashing the events of the week. It is always a good idea to be thinking before your session about what you want to put on the agenda, so you don't walk away feeling something important was missed.

In the first few sessions your therapist will model for you how to set agendas. So you don't have to instantly feel comfortable enough to add items to the agenda yourself, but can learn over time. This is itself a valuable process, helping you take charge of your problems and their solutions.

GOAL-SETTING IN CBT

Again, this is a collaborative process designed to maintain structure and focus. The point is to make the goals for your therapy those that are relevant to you, with input from your therapist to make sure they are clear and are what you actually want as opposed to what you think you should want. Goal-setting makes CBT productive by highlighting the possibility of change, making insurmountable problems appear more manageable, and increasing your hope of overcoming them.

While there are many different approaches to goal setting, one of the more common techniques used by CBT therapists is the SMART way.

SMART goal setting creates a clear and vivid picture of your goal and helps you maintain your motivation for achieving it.

This is what the acronym stands for :

Specific: Avoid generalisation. Be clear and focused

on exactly what you want. A specific goal has a much greater chance of being accomplished than a general goal.

Measurable: Establish concrete criteria for measuring progress toward the attainment of each goal you set. Ask yourself questions such as, "How much?" and "How many?" "How will I know when I have met my goal?"

Achievable: Make your goals attainable and feasible! How are you going to make this goal a reality? What can you do to make it more achievable?

Realistic: Is your goal realistic given your skills, time frame, etc? While setting high goals can be a great way of increasing motivation, it can also be disheartening when they are so high that you cannot reach them. This can leave you feeling like a failure.

Timely: Set realistic time-frames to avoid procrastination or giving up on your goal.

In pursuing your goals, if you find that that you are not able to complete a particular step along the path to your goal, take a closer look at it. It is possible that the step was too big, so do not be afraid to break it down further and go from there. It could also be that the step was not SMART from the start. See if you can rework it to make it one you can tackle.

Learning goal setting skills takes practice, but can be

invaluable in helping you do those things you have been avoiding or "just don't seem to get to" and feel a sense of accomplishment. Consider consulting with a CBT therapist if you think therapy may be helpful for you in reaching your goals.

Understanding the importance of Goal setting

A lot a people who come for therapy it is usually because their relationships are suffering, and so they are suffering. It could be a teenaged boy whose severe social anxiety prevents him from spending time with his friends, a woman with depression that makes it hard to be the partner she wants to be, a father whose expressions of anger have put distance between him and his kids, or the college student whose alcohol-fueled behavior has alienated her friends. It's hard for our relationships to thrive when we're hurting.

Effective therapy can improve our relationships, whether or not those relationships are the specific target of the treatment. A relatively brief course of cognitive behavioral therapy (CBT)—which has been tested more than any other treatment—can lead to marked benefits not only for the person in therapy but for those close to that person. Some of the major relationship benefits of CBT include:

1. Greater presence. It's hard to overstate the importance of our presence in a relationship, since we can't truly "relate" to someone who's not there. One of

the biggest complaints about partners that I hear in my practice is that "s/he isn't there for me". Sometimes the person means quite literally that their partner is absent—always traveling for work, for example. Just as often the problem is that even when the person is there in body, his or her mind is elsewhere.

Mindfulness-based CBT can address both of these issues; for example, training in mindfulness has been shown in multiple studies to increase one's ability to attend to the person we're with. A CBT framework can help translate one's intention to be present into a plan of action to make it happen.

Try it: The next time your partner talks with you about something, bring your full attention to what they're saying. Practice seeing the person as though for the first time, really focusing on them and what they're saying.

2. Less anxiety. When we're overwhelmed by anxiety, we're not our best selves. It's no surprise that untreated anxiety disorders take a toll on our closest relationships. For example, the need for a "safety companion" in panic disorder and agoraphobia can lead to strain as the supporting partner has to adjust his or her schedule to accommodate the other person's travels. Similarly the chronic worry in generalized anxiety disorder frequently leads to tension and irritability, causing conflict between partners.

For any anxiety diagnosis, the treatments with the most evidence for their effectiveness come from CBT. The relief a person feels from a marked reduction in anxiety extends to greater harmony in the relationship.

Try it: If you've struggled with uncontrollable anxiety, consider looking into the best treatments for your condition on the website for Division 12 of the American Psychological Association. You can also search the therapist directory of PsychologyToday.com for a therapist who provides the treatment you're looking for. If a suitable therapist is not available, you might pursue self-guided CBT (such as with this book), which has been shown to be effective.

3. Improved mood. As with anxiety, untreated depression wears on couples. It's a struggle to be the partner we're capable of being when we have no energy, little enthusiasm even for activities we would normally enjoy, and no sex drive, among other symptoms. After a typical course of CBT for depression—12 to 16 weeks—the average person will not only feel substantially better but will be able to function much more effectively and happier individuals make happier couples.

Try it: As with anxiety, you can search for a CBT therapist through the PsychologyToday.com therapist directory. There is also evidence that self-guided CBT can be effective for treating depression (e.g., CBT in 7 Weeks or Cognitive Behavioural Therapy), so CBT is available even when a therapist is not. You can even do Internet-based CBT (Be always careful for the Internet-based information)

4. Better sleep. As many as 23% of adults in the US suffered from bad sleep in the past month. When we're not sleeping well we tend to be cranky and impatient—not a recipe for the best interactions with the people who love us. Furthermore, insomnia can turn the bed into a place of worry and stress, which interferes with a cozy night's sleep beside our partner. CBT for insomnia (CBT-I) is typically 4 to 6 sessions and is the treatment of choice for insomnia. It helps a person fall asleep faster and sleep more soundly, and restores a strong association between the bed and sleep. And better sleep helps with pretty much everything.

Try it: If you've struggled with sleep, consult these guidelines for healthy sleep habits. The National Sleep Foundation has suggestions for finding a CBT-I therapist. CBT-I is also available through self-guided books (e.g., End the Insomnia Struggle) and apps.

5. Healthier relationship with alcohol. Problematic drinking can kill a relationship. Alcohol use disorder is tied to higher divorce rates, greater intimate partner violence, lower relationship satisfaction, and a host of

other problems. CBT can effectively target the thoughts and behaviors that maintain problems with alcohol, and replace drinking with healthier ways of coping. Interestingly, the treatment with the strongest research evidence is Behavioral Couples Therapy, with both the patient and his or her partner actively involved in the treatment.

For many individuals with an alcohol use disorder, lifelong abstinence is necessary. However, there is modest support for a treatment program that includes the possibility of moderate alcohol consumption for some people.

Try it: If you're interested in learning more about the Moderate Drinking program, some information are in this book (but not all the necessary ones so look for something more specific). You can also consult the PsychologyToday.com directory to find a therapist who provides Behavioral Couples Therapy for alcohol use disorders.

6. Happier kids. When a child is struggling with intense fears (e.g., phobias, obsessive-compulsive disorder), it can lead to tremendous stress for the family. Parents inevitably feel the strain when a child is refusing to go to school, struggling socially, or having problems at bedtime. As the saying goes, "You're only as happy as your least happy child."

Furthermore, most couples have somewhat different parenting styles, with one partner more lenient and the other more of the disciplinarian. A child's intense struggles will tend to amplify these differences, leading to conflict between the parents. At the end of the night when the kids are finally in bed and both parents just want to unwind, they may instead find themselves arguing about how best to help their child. Thus they may feel like their reserves are exhausted, with little left to give their child or each other.

The American Academy of Child and Adolescent Psychiatry recommends CBT as a first-line approach for treating many childhood conditions, including anxiety and OCD. Similarly, the American Academy of Sleep Medicine recommends behavioral treatments for sleep problems in infants and children, which can help both the child and the parents sleep through the night.

Try it: There are books available that detail how to apply the techniques of CBT to help your child with anxiety (e.g., Helping Your Anxious Child) or OCD (e.g., What to Do When Your Brain Gets Stuck). There are also many CBT-focused web resources available, such as from the International OCD Foundation and the Society of Clinical Child and Adolescent Psychology.

7. Healthier Thought Patterns. Even if we're not dealing with a diagnosable condition like anxiety, depression, insomnia, or a substance use disorder, the tools of CBT can have powerful effects on our

relationships. CBT is based on an understanding of the connections among thoughts, feelings, and behaviors. When our thought patterns are aligned with reality, they generally lead to positive feelings and behaviors. However, when our thoughts become distorted in some way, they start to work against us, including in our relationships.

For example, we might notice that our partner left his clothes on the floor and think, "He expects me to pick up after him. He thinks I'm his maid." The result might include a fight driven by resentment and defensiveness. Or we could think that our partner seems distant and tell ourselves, "She's unhappy with me and our relationship," leading us to withdraw in turn.

The cognitive part of CBT encourages us first of all to notice the thoughts we're telling ourselves; oftentimes they happen so quickly and automatically that we don't even recognize the story our mind is creating. Once we've identified the thoughts we can test them out to see if they're accurate. Maybe our partner's clothes on the floor say nothing about his view of us, or expectations that we clean up his mess. And perhaps our partner's preoccupation has nothing to do with our relationship and everything to do with her worries about her ailing mother. With practice we can replace distorted and destructive thoughts with more accurate and constructive ones.

(Importantly, cognitive techniques are not about fooling ourselves or pretending things are better than they are. It would be important to know if our thoughts are actually valid so we can deal with the situation directly.)

Try it: The next time you're upset with your partner, write down the thoughts you notice and the emotions you feel. Then ask yourself the following questions (adapted from Retrain Your Brain: Cognitive Behavioral Therapy in 7 Weeks):

What is the evidence for my thought?

Is there evidence that contradicts my thought?

Based on the evidence, how accurate is the thought?

How could I modify this thought to make it fit better with reality?

8. Greater enactment of our intentions. All of us want to be the best significant other we can be. We want to be attentive, supportive, generous, patient. And like anything else, the road to impoverished relationships is paved with the best of intentions. If we're not deliberate about living out our values, we risk leaving them in the abstract—vague platitudes without substance.

For example, we might tell ourselves, "My family matters to me more than anything," and then live as though family is our last priority. We might idealize presence in our relationship yet attend more to our phone than to those around us.

The tools of CBT can help even when there is no "disorder"—when we simply want to commit to sustained action that supports our deepest values. It starts with taking an inventory of our relationship, and setting clearly defined goals based on what we find—for example, To listen when my spouse talks to me. It includes identifying specific behaviors we want to practice that will move us toward our goals.

We might plan, for instance, to turn off our phone during dinner and focus on our conversation. The goals and activities can be anything that's important to us in our relationship—we get to decide. It can be very beneficial to collaborate with our partner in the process by asking what they need more of from us.

A CBT approach also includes planning activities as specifically as possible so they stay on our radar, like putting "Come home early to make dinner" in our calendar, and protecting the time. Through daily practice of our intentions, we give our relationship the nourishment it needs not only to last but to be extraordinary.

CHAPTER FOUR

CBT FAQS

Answers to frequently asked questions about CBT, such as how to find a therapist and what to expect in treatment, are outlined below.

How will I know if CBT is for me?

Although many people can benefit from CBT, not everyone finds it helpful. You might find that it just doesn't suit you, or doesn't meet your needs.

I think it can go either way. I've been through a long course of CBT a few years ago, and for some people it helped, whereas for others it didn't.

Before deciding to have CBT, it might be helpful to think about the following:

Is short-term therapy right for me? If you have severe or complex problems, you may find a short-term therapy like CBT is less helpful. Sometimes, therapy may need to go on for longer to cover fully the number of problems you have, and the length of time they've been around.

Am I comfortable thinking about my feelings? CBT can involve becoming aware of your anxieties and emotions. Initially, you may find this process uncomfortable or distressing.

How much time do I want to spend? CBT can involve exercises for you to do outside of your sessions with a therapist. You may find this means you need to commit your own time to complete the work over the course of treatment, and afterwards.

Do I have a clear problem to solve? You may find CBT is less suitable if you feel generally unhappy or unfulfilled, but don't have troubling symptoms or a particular aspect of your life you want to work on.

Most people know within the first few sessions if they are comfortable with CBT and whether it is meeting their treatment needs. The therapist will also check to see that CBT is the right "fit" for you. When the fit is not quite right, the therapist may adjust the treatment or suggest other treatment options. However, in general, CBT may be a good therapy option for you if:

You are interested in learning practical skills to manage your present, day-to-day life and associated emotional difficulties.

You are willing and interested in practising change strategies ("homework") between sessions to consolidate improvement.

CBT may not be for you if you want to focus exclusively on past issues, if you want supportive counselling, or if you are not willing to do homework between sessions.

How can I find a qualified CBT therapist in my area?

It is most probably wise when seeking a therapist who provides CBT to try and assess which of the above-mentioned categories they may fall into. Many therapists will report that they use CBT, but few have actually had specific supervised training in the model and many do not provide cutting-edge treatments for specific disorders.

In addition, it would also be important to check whether or not the therapist has specific experience in your area of concern. For instance, an individual may have plenty of experience in working with anxiety or depression, but little experience or expertise in treating tic disorders, insomnia or psychosis, for example. Below is a list of reasonable questions that one could ask a therapist prior to making an appointment or during an initial appointment aimed at assessing their experience with CBT.

Have you received any training in CBT?

Training could vary from an introduction during a masters programme to an internationally-certified qualification.

Did your training include individual supervision of case material from an experienced or internationally-certified CBT therapist?

It is important to distinguish between theoretical training and practical, supervised training that includes individual supervision in the use of CBT in working with clients.

Does your CBT training include international certification from an international training standards committee?

Typical international certification is from the Beck Institute or the Albert Ellis Institute. Other training centres in the USA and the UK also exist.

How many years' experience do you have in practising CBT?

We all start with little experience and build our knowledge base. It would be wise not to doubt a well-trained and supervised CBT therapist with one or two years' experience. It, however, goes without saying that those therapists with greater practical experience bring the expertise accumulated from this experience to helping you.

In what way do you use CBT in your practice?

Many people will tell you that they integrate CBT in to their practice. This typically would entail using skills or techniques from more than one theoretical orientation at the same time. This approach is known as eclecticism. This is not unethical nor an unacceptable practice. Should you look for a CBT therapist specifically, this is obviously not the type of service provided by a therapist with an eclectic approach. Therapists will also tell you that they use different theoretical approaches to treat different clients and problems. This is also an acceptable approach. It would, however, be important to determine if the therapist is sufficiently trained in CBT and if the therapist would suggest CBT for your specific problem.

Ask if the therapist has training in the CBT treatment of your specific disorder.

It important to know that the therapist has disorder-specific training in the condition that you require help with.

To find the contact information for certified CBT therapists in Canada and the United States, consult the following resources:

Academy of Cognitive Therapies
www.academyofct.org

Association for Behaviour and Cognitive Therapies
www.abct.org

What can I expect on my first visit with a CBT therapist?

You will meet your therapist and they will introduce themselves, ask for your preferred name and help you to feel at ease. They may also introduce a little bit about what CBT is and the way that the sessions will be. You do not need to know anything about CBT before you begin your therapy and your therapist should explain it in an easy-to-understand way.

Your therapist will talk to you about confidentiality, the number of sessions you may work towards, how often your sessions will be, how long they will last, how to cancel appointments and what happens if you miss an appointment.

Your therapist will usually have the information you gave at your initial screening appointment but may ask you a little more about the main problem you would like some help with and about you and your circumstances. This is not to get you to 'tell your story all over again', but to help you and your therapist reach an understanding of how you will work together and what you will focus on within the sessions.

Your therapist may also ask you to think of some positive goals so that you have something you are working towards. This can help you to connect with something important to you or to work towards something you would like to do if you no longer had the problem.

Your therapist may talk to you about 'homework'. Homework is an integral part of CBT and after each session you will think with your therapist about how you can apply the things you have done in the session to your day-to-day life. After your first appointment, you may be asked to keep a monitoring diary or to read some information/handouts relevant to your problem.

You will do the same questionnaires you completed at your first appointment. These will be asked at every appointment to help you and your therapist keep track of your progress. You may sometimes be asked to do an additional questionnaire more specific to your problem, however, this will vary from client to client.

Your therapist will check with you if you are at any risk to yourself or from anyone else. This is something we ask every client so that we can help you to stay safe.

You will make your next appointment with your therapist.

Your therapist will go at a pace that suits you and the above points are a guide only. You may do more or less than this in your first appointment and that is OK. You can also ask questions and express any concerns about therapy at any time in your appointment. Your therapist wants to work with you as a together to help with your recovery. While your therapist can offer ideas about what has helped others with similar problems in the past, they also know that you are the expert in your problem and how it affects you, and you should both be playing an active role in therapy together.

What happens in a CBT session?

CBT sessions can be given to clients individually or in groups. Both formats follow the same predictable structure, as follows:

Mood check: The therapist asks about your mood since the previous session. This may include the use of scales to assess depression, anxiety or other emotional problems. The purpose of the mood check is to see if your mood improves from session to session.

Bridge: The focus of the previous session is reviewed to create a bridge to the current session.

Agenda: The therapist and client identify issues to address in the current session that will act as the agenda.

Homework review: Homework from the previous session is reviewed to note progress and troubleshoot any difficulties that may have emerged.

Agenda items: Agenda issues are addressed using cognitive and behavioural strategies.

New homework: Exercises and tasks for the upcoming week are assigned.

Summary and client feedback: Wrap up of the session.

How can I get the most out of CBT between sessions?

CBT is a treatment approach that teaches you skills to become your own therapist over time. You learn new skills in the therapy sessions, but ultimately much of the change occurs between therapy sessions when practising the skills in your own environment as part of homework. Early in treatment, the CBT therapist will suggest homework, such as monitoring thoughts and behaviours, taking steps to reduce avoidance behaviour, conducting experiments to test out predictions and completing worksheets to challenge negative thoughts or beliefs. As treatment progresses, you will learn to set your own homework between sessions to help you accomplish your treatment goals. Research demonstrates that the more you successfully practise the skills of CBT in your homework, the better the treatment outcome.

How frequent are the sessions?

CBT usually starts out with weekly sessions. As treatment progresses, sessions may be spaced further apart, such as every two weeks or month. Once people have finished a course of CBT, it is common for them to return for occasional "booster" sessions to keep up their progress, deal with any setbacks and prevent relapse of problems. Again, hospital-based programs are typically pre-set (e.g., a group meeting weekly at the same time for 12 weeks) and are less flexible in terms of scheduling and spacing sessions than community-based CBT therapists.

Do I need to prepare for CBT sessions?

Preparing to discuss a specific problem at each session helps you to get the most out of CBT. Coming prepared helps to keep you focused on your goals for therapy. It also helps to build a therapeutic relationship between you and your therapist and to communicate well throughout the session.

Will the CBT therapist be able to understand and appreciate my own unique background?

Research on CBT demonstrates that it is an effective treatment regardless of gender, race, ethnicity, culture, sexual orientation or social economic status. CBT therapists are trained to recognize the importance of cultural values and to adapt their treatments to meet culturally unique needs. They are trained, for example, to:

Be aware of their own personal values and biases and how these may influence their relationship with the client.

Use skills and intervention strategies that are culturally appropriate for the person being seen.

Be aware of how certain cultural processes may influence the relationship between the therapist and client.

As a client in CBT you should feel that you can openly discuss aspects of your culture or sexual orientation, for example, and that your treatment will be delivered in a manner that is consistent with these values.

Is CBT an effective treatment for children and adolescents?

CBT has been adapted for use with children and has been shown in research to be an effective intervention for a variety of clinical problems that can emerge in childhood, including anxiety and depression. The content and pacing of the therapy is adjusted to be appropriate for the child's level of development. Often, the CBT therapist will work with the parent and child—the younger the child, the more involved the parent will be in learning and delivering CBT strategies for their child's problem.

What are the common barriers that come up in CBT?

Barriers to treatment can include:

- Perceived stigma associated with mental health treatment
- Difficulty identifying and distinguishing emotions and their intensity
- Difficulty in reflecting on thoughts
- Difficulty tolerating heightened emotions
- Not completing homework
- Financial constraints
- Chronic conditions and multiple difficulties
- Low optimism toward improving
- Avoiding treatment sessions.

The therapist will work with you to reduce these barriers and will also offer strategies that you can use to overcome barriers.

Should I start treatment with medications or CBT or both in combination?

Many people who seek treatment for emotional difficulties in Canada are first treated by their family doctor with one or more medications (e.g., antidepressants) before CBT is considered. Not much research has been done to show whether it is best to start treatment with medication or CBT or both. However, research has shown that CBT, with or without certain types of medications, is equally effective for the treatment of anxiety problems, and that CBT and medications together are best for treating severe depression and psychosis. Importantly, taking these kinds of medications has not been shown to interfere with CBT, and in some instances may help people to get more out of the therapy.

There is, however, one class of medication commonly prescribed to people with anxiety problems, known as benzodiazepines, which can potentially limit the benefits of CBT. Medications in this class include clonazepam (Rivotril), alprazolam (Xanax) and lorazepam (Ativan). While these medications can rapidly relieve and control anxiety in the short term, they can also make it harder to learn new things, which is essential to benefit from CBT and reduce anxiety in the long run. If you are taking these medications and are about to start CBT, your CBT therapist will want to review the advantages and disadvantages of continuing with these medications during CBT treatment.

When considering or starting CBT, discuss all of your questions or concerns about your medications with your CBT therapist and/ or your prescribing medical doctor. Medications can be monitored and discussed throughout treatment and adjusted depending on your progress in CBT. Note that recommendations or changes to your medication can only be made by your prescribing doctor. If your CBT therapist is not a doctor, he or she can—with your permission—communicate with your doctor to help ensure that you receive the optimal combination of treatments.

How can I stay well after finishing CBT?

A major goal of CBT is for you to become your own therapist and to continue to practise CBT skills even after you are feeling better. You may also wish to return for follow-up or "booster" sessions from time to time. A key component of CBT treatment is teaching relapse prevention strategies. This includes helping you learn to identify the triggers and early signs of relapse and to develop an action plan to prevent downward spirals of negative emotions.

CHAPTER FIVE

ALTERNATIVE COGNITIVE BEHAVIOURAL APPROACHES

Of the hundreds of psychological treatments available, several are closely related to CBT, but have distinct approaches. Although the effectiveness of these alternative CBT approaches is not as well proven as the mainstream approach described so far in this guide, the four introduced on the following pages have been found to be effective in helping people with certain types of problems. Overall, they all share a common goal of helping people learn how to "let go" of focusing on and reacting to their thoughts.

Mindfulness therapy and mindfulness based cognitive therapy

Mindfulness techniques can be used to help people distance themselves from their negative thinking and recognize that thoughts do not have to determine behaviours. Mindfulness is a state of awareness, openness and receptiveness that allows people to engage fully in what they are doing at any given moment. Mindfulness skills are mainly taught through meditation; however, other experiential exercises (e.g., walking or eating with awareness) can also be used to teach these skills.

A middle aged woman is feeling overwhelmed, anxious, depressed, hopeless and depleted of all of her energy. She explains to her doctor that she has been having feelings of self doubt and is desperate to turn back the clock to a time in her twenties and thirties when she knew she was "beautiful", "thin", and "revered". She has tried everything from Pilates, energy healing, life coaching, herbal remedies, hypnotherapy, Reiki, and has visited an array of holistic spas and retreats.

Yet she continues to fee chronic emptiness and impending doom. It is as if she is suffering from a should sickness. Not doubt the alternative treatments were effective, at least initially. But the days of long-lost youth and feelings of worthlessness returned. When her psychotherapist recommends learning mindfulness meditations, she agrees in her desperation.

Psychotherapy and pharmacological solutions are some of the common treatment options for patients with mental disorders. But, alternative treatments for psychiatric conditions are increasingly becoming popular. Lately, doctors have been advising alternative treatments, like meditation and yoga, for mental issues, particularly for anxiety disorders and depression.

According to a recent study published in the Journal of Child and Adolescent Psychopharmacology, significant changes were evident in the brain regions that control emotional processing of youths who were given mindfulness-based therapies.

Though anxiety disorders are common among children and adolescents, antidepressants administered to treat the condition are many a times not tolerated well by children who are at a high risk of developing bipolar disorder.

So, the researchers at the University of Cincinnati (UC) have now found out how cognitive therapy that utilizes mindfulness techniques, such as meditation, quiet reflection, and facilitator-led discussion, may help as an adjunct to pharmacological interventions. The study was part of a larger investigation to understand the effectiveness of mindfulness-based therapy.

The respondents were chosen from a group of youths who had anxiety disorders (generalized, social and/or separation anxiety) and who have a parent with bipolar disorder. The study, published in the Journal of Child and Adolescent Psychopharmacology in July 2016, tried to evaluate neurophysiology of mindfulness-based cognitive therapy in children who had higher risks of developing bipolar disorder.

Mindfulness therapy increases activity in brain

The nine participants aged 9-16 years were made to undergo functional magnetic resonance imaging (fMRI) while they were involved in a continuous performance of tasks with emotional and neutral distractors prior to and following 12 weeks of mindfulness-based cognitive therapy.

"Our preliminary observation that the mindfulness therapy increases activity in the part of the brain known as the cingulate, which processes cognitive and emotional information, is noteworthy," said co-principal researcher of the study Jeffrey Strawn, M.D., an associate professor in the Department of Psychiatry and Behavioral Neuroscience, and director of the Anxiety Disorders Research Program.

"This study, taken together with previous research, raises the possibility that treatment-related increases in brain activity (of the anterior cingulate cortex) during emotional processing may improve emotional processing in anxious youth who are at risk for developing bipolar disorder," he added.

Speaking about the effectiveness of mindfulness techniques, co-author of the study Sian Cotton, Ph.D., an associate professor of family and community medicine at UC, said, "Mindfulness-based therapeutic interventions promote the use of meditative practices to increase present-moment awareness of conscious thoughts, feelings and body sensations in an effort to manage negative experiences more effectively." These alternative approaches augment traditional treatments offering new strategies for coping with psychological problems, he said.

Recovery roadmap

The researchers noted that increases in mindfulness were associated with decreases in anxiety in the participants.

However, the researchers called for further studies into this for more clarity. "The path from an initial understanding of the effects of psychotherapy on brain activity to the identification of markers of treatment response is a challenging one, and will require additional studies of specific aspects of emotional processing circuits," Strawn said.

For any mental health condition, be it anxiety disorder or depression, early intervention is the key. Hence, if a loved one is exhibiting any psychiatric symptom, seek the advice of a doctor immediately.

Mindfulness skills can be broken down into three categories:

Defusion: distancing oneself from and letting go of unhelpful thoughts, beliefs and memories.

Acceptance: accepting thoughts and feelings without judgment, simply allowing them to come and go rather than trying to push them out of awareness or make sense of them.

Contact in the present moment: engaging fully in the here-and-now with an attitude of openness and curiosity.

Mindfulness skills promote freedom from the tendency to get drawn into automatic negative reactions to thoughts and feelings.

Mindfulness techniques have been used in the treatment of chronic pain, hypertension, heart disease, cancer, gastrointestinal disorders, eating disorders, anxiety disorders and substance use disorders. Mindfulness-based cognitive therapy has been found to be effective in reducing relapse to depression.

Acceptance and commitment therapy

While some therapies attempt to change upsetting thoughts and feelings, acceptance and commitment therapy (act) helps people to simply notice and accept thoughts and feelings in the present moment.

ACT views psychological suffering as being caused by avoiding or evaluating thoughts and feelings, which in turn can lead to ways of thinking that interfere with our ability to act consistently with important personal values. The focus of act is on helping people accept what is out of their personal control while committing to doing what is within their control to improve their quality of life.

ACT aims to help people handle the pain and stress that life inevitably brings and to create a rich, full and meaningful life. People learn how to deal with painful thoughts and feelings in ways that have less impact and influence over their lives. For example, they learn to:

- Distance themselves from upsetting thoughts (cognitive defusion)
- Accept experiences in the present moment
- Discover important and meaningful personal values
- Set goals consistent with these values
- Commit to take action

ACT has been found to be effective in treating depression, anxiety, stress, chronic pain and substance use disorders.

Dialectical behavioural therapy

Dialectical behavioural therapy (dbt) is an effective treatment for people with excessive mood swings, self-harming behaviour and other interpersonal problems related to the expression of anger.

DBT has an individual and a group component. In individual therapy, the therapist and client follow a treatment target hierarchy to guide their discussion of issues that come up between weekly sessions. First priority is given to self-harming and suicidal behaviours, then to behaviours that interfere with therapy, and next to improving the client's quality of life. The quality-of-life improvement part of the therapy involves identifying skills that the person has but is not using to full advantage, to teaching new skills, and to discussing obstacles for using those skills. Weekly group therapy focuses on acquiring new skills.

Clients keep diary cards to help monitor their use of the skills and have access to 24-hour phone consultation with their therapist.

The four modules of dbt are core mindfulness, emotion regulation (e.g., identifying and labelling emotions and reducing vulnerability to negative emotions), interpersonal effectiveness (e.g., assertiveness skills) and distress tolerance skills (e.g., crisis survival skills such as distracting, self-soothing and improving the moment).

Meta-cognitive therapy

Metacognitive therapy is sometimes described as a type of therapy that involves changing how people think rather than what they are thinking about. In this way, metacognitive therapy is distinct from cognitive behavioral therapy, which focuses more on the content of people's thoughts.

Two psychologists, Dr. Adrian Wells of the University of Manchester and Dr. Gerald Matthews of the University of Cincinnati, developed the theory underlying metacognitive therapy in the early 1990s. Initially intended only for patients with generalized anxiety disorder, metacognitive therapy has since been adapted for use in treating a variety of mental health problems.

According to metacognitive theory, maladaptive thinking that occurs in various psychiatric disorders tends to take on a life of its own, growing from thoughts about a specific situation to a more global world view. For example, people who are anxious may initially worry about external situations, such as missing a train. With time, however, they may begin to develop a second type of worry, focused on their own thought processes. In essence, according to this theory, they begin to worry about being worried.

In people with attention deficit hyperactivity disorder (ADHD), problems with metacognition more often encompass difficulty in planning or executing tasks. The goal of metacognitive therapy in ADHD is to improve organization skills, planning, and time management.

Most of the research evaluating metacognitive therapy has focused on anxiety disorders such as social phobia, post-traumatic stress disorder, and generalized anxiety disorder. Only a few studies have evaluated its use in helping adults with ADHD (none have been conducted in children). The research is therefore preliminary, but so far promising.

In one study, researchers at the Mount Sinai School of Medicine recruited 88 adult participants with ADHD, diagnosed using a structured interview based on criteria in the Diagnostic and Statistical Manual of Mental Disorders, Fourth Edition (DSM-IV). The group was carefully selected; few had substance abuse or other types of coexisting psychiatric disorders, which is more the norm in the community. The researchers randomly assigned half of the participants to metacognitive therapy and the others to supportive therapy, both offered in group formats consisting of 12 weekly sessions.

Those assigned to metacognitive therapy underwent a sequence of sessions designed to begin with learning specific skill sets (such as using a daily planner) and then progressing to broader abilities (organizing and executing a complex project). Those assigned to supportive therapy received encouragement and reinforcement of productive behaviors.

Some participants improved after both interventions, but those assigned to metacognitive therapy showed a greater degree of improvement on both objectively rated and self-perceived measures of organization, ability to complete tasks, and other practical components of attention skill. A greater proportion of people assigned to metacognitive therapy responded to therapy, defined as at least a 30% improvement in ADHD symptoms. In all, 19 of 41 (42%) of participants who completed metacognitive therapy responded, compared with five of 40 (12%) of those who completed supportive therapy.

This suggests that metacognitive therapy may indeed help some adult patients with ADHD. But before your husband tries it, just be aware that there are no long-term data about whether the potential benefits are retained over time. Your husband may also find it helpful to investigate other options to improve organization skills, such as cognitive behavioral therapy or coaching.

Meta-cognitive therapy (mct) was first developed to treat generalized anxiety disorder and is now also used to treat other anxiety disorders and depression.

Metacognition is the aspect of cognition that controls mental processes and thinking. Most people have some direct conscious experience of metacognition. For example, when a name is on the "tip of your tongue," metacognition is working to inform you that the information is somewhere in memory, even though you are unable to remember it.

People with depression or anxiety often feel as though they have lost control over their thoughts and behaviours. Their thinking and attention become fixed in patterns of brooding and dwelling on themselves and on threatening information. They develop coping behaviours that they believe are helpful, but that can actually worsen and prolong emotional distress. This pattern of thinking is called cognitive-attentional syndrome (cas).

In mct, people learn to reduce the cas by developing new ways of controlling their thinking and attention and of relating to depressive or anxious thoughts and beliefs. They also learn to modify the beliefs that give rise to the cas.

CHAPTER 6

APPLYING COGNITIVE BEHAVIORAL THERAPY (CBT) TO OVER-THINK NEGATIVE PATTERNS

Negative thinking can slow down the recovery of depression, and the reason is clear: If you think negative thoughts, you are more likely to remain depressed. But the way people with depression deal with positive emotions is less obvious. Researchers made a surprising observation: People with depression don't lack positive emotions, they just can't feel them. This cognitive style is called "dampening," it involves suppressing positive emotions with thoughts such as "I don't deserve to be happy" or "This good feeling won't last." For example, a new mother with postpartum depression may say she doesn't deserve to recover because she's a bad mother because she's depressed. Why do depressed people think so? We refers to this negative voice as a defensive pessimism-protection against high hopes. "you don't want to be foolish, so you use positive thoughts to protect yourself from possible disappointment."

How CBT can help with negative depression

Depression therapy has been found to be significantly helpful in treating depression. In CBT, you work together with your therapist to agree on behavior patterns that need to be changed. The objective is to recalibrate the part of your brain that holds happy thoughts so tightly.

"The root of the dampening effect could be an unexpected reaction to a major life event" through CBT, you and your therapist address it and work to put it into perspective." Regular CBT sessions and work you do on your own outside of therapy can help strengthen the new patterns: "It can be very liberating to be able to recognize these negative thoughts and leave them behind,"

CBT Techniques to Counteract the Negative Thinking of Depression

It was discovered that people with depression rarely respond well to self-study by CBT techniques to counteract negative thinking about depression. For this reason, we recommend that at least six weeks be committed to CBT. Your therapist will teach you CBT strategies to counteract depression-related negative thinking. She or he can also help you keep track of the techniques. You may end up working on five CBT strategies with your therapist:

Locate the problem and brainstorm solutions.

Journaling with your therapist and talking can help you discover the root of your depression. Once you have an idea, write down exactly what disturbs you in a simple phrase and think about ways to improve the problem. A hallmark of depression is despair-an unbelief that things can ever get better. Writing a list of things you can do to improve a situation can help ease the feelings of depression. If you are fighting for loneliness, for example, action steps to try may include joining a local club based on your interests or signing up for online dating. Write self-declarations to counter negative thoughts. After finding your depression's root problems, think about the negative thoughts you use to dampen positive thoughts.

Write a self-declaration to counteract any negative thinking.

Remember your self-statements and repeat them to yourself when you see the little voice creeping in your head to snuff out a positive thought. You will create new associations in time to replace negative thoughts with positive ones.

The self-declaration should not be too far from negative thinking, or the mind may not accept it. For example, if the negative thought is, "I'm so depressed right now," instead of saying, "I'm really happy now," it could be better to say, "Every life has ups and downs, and mine also has ups and downs." At the same time, your mind applauds the fact that joy is kept in check to protect against disappointment. "Recognizing that part of you is trying to do something healthy is okay."

Self-reporting is sometimes too routine and needs to be refreshed. We recommend that you translate your self-declarations into other languages you can speak or rephrase, possibly even bumping up your joyful feelings. "For example, the self-statement" it's okay to explore my ups" may turn out to be "it's okay to have a super "up" day."

Find new opportunities to think positive thoughts.

People who enter a room and immediately think, "I hate the color of the wall," could instead train to locate five things in the room they feel positive as quickly as possible. Set your phone to remind you of something positive three times a day. It recommends that someone else work on the same technique should "buddy up." You and your friend can be excited to share positive thoughts and experiences throughout the day.

Complete every day by visualizing the best parts.

At the end of each day, write down the things you're most grateful for in your life or type them into an online journal. The recording of positive thoughts and even sharing them online can help you form new associations in your mind or create new paths. Someone who created a new way of thinking could wake up in the morning and think, "Ugh, another working day "to "What a beautiful day it is."

Learn to accept disappointment as a normal part of life.

Deceptive situations are part of life and your response can affect how fast you can move forward. Someone who goes through a breakup may blame him or herself or even gain weight, saying, "What's the point of looking good? I'll never meet anyone else. "Maybe a better approach is to be disappointed and remember that some things are out of your control. Work on what is in your control: Write down what happened, what you learned from the experience and what you can do next time differently, taking care of overly negative thoughts. This can help you to move forward and feel better about your future.

Cognitive behavioral therapy Effective tips

Research has shown that we can have better control of thoughts and feelings by identifying our distorted thoughts and beliefs. Having distorted thoughts or beliefs doesn't mean we're wrong. At different times in our lives, we have all distorted thoughts and beliefs. Examples of distorted thoughts:

Over-generalization

Sometimes we can see things as everything or nothing. If one thing goes wrong with a project, for example, we may think the whole project is a failure.

Or, if there's one thing that upsets us about a person, we can decide that we don't care about him.

Mind Reading

We assume we know what someone thinks. We can say that someone thinks we're "stupid "or doesn't like us, even if there's no evidence to support this thought. That's called reading the mind. We exaggerate how "abhorrent "something is or imagine the worst outcome possible. Maybe our boss wants to talk to us and we're disastrous to be fired. Or, it's raining on one of the holidays and we think "this is the worst thing that could have happened."

Fortune Telling

We think we surely know what will happen. We say, for example, "I know I won't get this promotion "or "I won't be able to handle this task." Specific behaviors or skills, including social skills, assertiveness, organizational skills and relaxation techniques, are also taught. During and between sessions, these are taught. Below, there are seven pearls I will share with you that I have found useful in my practice over the years:

Discuss treatment goals

During the initial evaluation phase, it is important to work together on treatment goals. This helps to maintain focused and productive treatment. Therapy can end up focusing on any problem that arises this week without objectives and can interfere with the progress of the original problems presented. Sometimes the patient may not be able to describe a goal specifically except for a vague "I want to be less anxious "or "I want to be happier." It's okay at the start. However, you should return to this discussion about goals in the first couple of months to see if they can be described in more specific terms. For example, if someone has depression, the goals may include: Finding a fuller job, going back to college, exercising three times a week, making two new friends and stopping marijuana use.

Each session with an agenda

Each session should begin with a collaborative agenda between the therapist and the patient. Again, this helps to focus and make the session more efficient. The agenda should include follow-up on previous session homework, check-in on mood and week, bridging or reviewing the topics and progress from the previous session, and topics related to the current session, which are related to a specific objective.

Discuss where to address the issue

Most therapeutic objectives will have several components, including distorted thoughts, beliefs or behaviours. Therefore, during the session, decide collaboratively on which level to meet the objectives. If you work on distorted thoughts, it is important to determine which thoughts or images lead to distress, such as anxiety, low mood or a certain behavior. If you work on certain behaviors such as social skills or relationship problems, it is important to discuss when the skills are used and how likely they are used. Another useful technique for addressing behaviors is the play of roles and visualization that helps to practice skills and address any behavioral blocks or anxieties.

Use Flashcards

Flashcards can be used to remember the session's key points or a mantra that can help with certain thoughts or feelings. If I work with a patient who is struggling with depression, I will name the flashcard something like "Survival Kit "and it will include strategies to cope with depression, such as reaching a friend, leaving the house, reaching me or taking care of a small chore.

Stay focused

At the start of treatment, therapeutic objectives are discussed. The therapy session can sometimes go in a direction unrelated to any of the treatment goals. This is appropriate at certain times, but if this happens every session and for the whole duration, the progress of therapy may be limited. In CBT, structure is important, but flexibility is also important. This would be a time to work together to discuss whether to continue with the current diversion or issue being discussed or to return to what has been discussed in the agenda.

Assign homework

A collaborative discussion on homework or "action tasks "takes place between sessions towards the end of each session. If one of the problems is time management or the recording of thoughts and images that occur during stressful periods in a notebook to discuss and address at the following session, an action task may be to buy a calendar. Always ensure that the homework or action task is monitored at the next session, or it creates the impression that working on problems or objectives between sessions is not a crucial part of improving.

Ask for feedback

At the end of the session, ask what was going well during the session, what could have gone better and what are the main messages to take away. This helps build the alliance, improve sessions in the future and maximize progress.

Cognitive behavioral therapy is a highly effective form of therapy with or without medicines and an excellent way to practice psychiatry.

Cognitive behavioral therapy at home Tips that can help ease your anxieties

CBT is based on the idea that our thoughts, emotions and behaviors are interconnected and that one can change the other. This may sound trendy, but it is also effective and has been studied rigorously. CBT varies from anxiety to depression to schizophrenia to substance use disorders for all types of mental health problems.

The goal is to learn skills to address real-life problems outside the therapist's office, Lindgren says. The more you practice, the more a habit will become of CBT skills.

"If you're someone who has good intentions but needs someone to be accountable, I'd meet a therapist" but if you know you're a good self-taught person, it's reasonable to think about doing it on your own."

Here are her tips for practicing at home (or wherever you are).

Change your perspective

You can modify problem thoughts by using a technique called cognitive restructuring, which can help you change your behaviour. The next time you notice that you are anxious or depressed, ask yourself: What do I think about or what emotions do I struggle to make me feel like this? Notice if certain thoughts or memories cause distressing physical symptoms; you can even list them. This will help you understand how your emotions and thoughts are connected and what is triggering you.

Balance your thoughts

Many struggles in mental health involve thoughts or predictions that influence behavior that are distressing but inherently faulty. For example, if you're anxious when you're in crowds and actively avoid them, you might be told that if you tried to go to a crowded place-like a sports game or concert-you'd panic, do something to embarrass yourself, and you wouldn't enjoy it. This belief then strengthens your avoidance. Notice how your brain makes decisions based on fear or avoidance and then ask yourself: What is the evidence for this thought? Are there any cold, hard facts that things are going to go wrong, or do I just speculate? Consider if you could have other thoughts that would be more balanced or helpful. If you changed your thinking a little bit to be less frightening or negative, what new emotions could arise? If you work to balance your thoughts, your emotions and behaviors will probably follow.

Be patient with yourself

Change won't happen overnight, so don't expect it if you try CBT alone (or even with a therapist to guide you). Instead, your goal should be to develop your skills so that you feel better equipped to deal with any challenges your mental health wants to take. Focus on getting ready for small victories, then build up your goals slowly over time. Be proud of any positive change, however small it may seem. Recognize that progress is not linear; it will be easier for some weeks, harder for others, and that's normal.

Be kind to yourself

Without even realizing it, it's easy to get caught up in negative self-talk. But getting constantly on yourself won't inspire the confidence you need to make you feel better. When you notice negative thoughts-things like "Why can't I get it together?" or "other people have no problem "-replace them with something nicer. Ask yourself if your friends will ever tell you the things you say to yourself. No, no? So don't let yourself tell them either. This doesn't mean you should apologize for yourself if you made a mistake or did something wrong, but instead encourage you to cut the slack you usually reserve for others.

Do what you love

Anxiety, depression and other mental health struggles can remove the activities that matter to you in life, either because you are afraid of them or because you have no motivation to pursue them once. Maybe you liked reading, but now you're tired. Or you may have liked to go out with your friends, but now you're afraid to be away at night. Take the time to do one or two things regularly that always brought you joy and did your best to be present instead of being distracted or worried about the past. Then ask yourself how you feel you've done it. Did you feel better?

Be Self-aware

Maybe when you try to fall asleep or beat yourself over something you told a friend when you should finish an important work project, you're ruminating about work problems; either way, you're not focused on the moment. Instead, try switching your thoughts whenever they're not in line with what's going on right now. Ask yourself: Do my emotions reflect what's happening right now? If not, concentrate on the senses. What are you seeing and hearing? What's happening around you in the world? Instead of what happened in the past or what you're afraid will happen in the future, try to be aware of what's right before you.

A bright future

In the end, one of CBT's most powerful things is that it can give you hope.

"Inherently, it's optimistic. It teaches you to believe that change is possible and that you can change your life."

CHAPTER 7

MOST COMMON ERRORS MADE IN COGNITIVE-BEHAVIORAL THERAPY

In the course of my clinical practice, I have found a number of common errors that can prevent cognitive-behavioral therapy (CBT) from being as effective as it could be. The following describes these errors and how to correct them.

ERROR 1: Not Understanding the Necessity of Repetition for Change

A few years back when I was lecturing a group of psychiatric residents (future psychiatrists) about using CBT with sexually traumatized clients, one of the doctors stated "I tried that cognitive therapy with my patients and it doesn't work at all!" As I explored with her the specifics of what she had tried with her patients, I ascertained that, basically, she had told them how they should think and then expected them to change their thinking.

When I asked her what method she used to help them practice and repeat the new ways of thinking in daily life, she was speechless. I then explained that the key to successful cognitive therapy was repetition of the statements that challenge the irrational thinking and that the therapist must use a method to help clients do that.

People don't change their thinking just because they are told to think differently. If that was the case, most people wouldn't need CBT because someone else has probably already told them how they should think. The difference with CBT is that a variety of methods have been developed to provide the steps to change thinking.

The importance of repetition as a key component of CBT can't be emphasized enough. This is true of any new skill. When I was training for my black belt in Kenpo karate my instructor told me "When you have done this self-defense move 6,000 times, it will be automatic." And he was right! At first, I was awkward and slow in executing the movements. I had to think about every aspect of each move. I thought initially that it would never be automatic. However, I continued to practice. One day (after years of training) when I was teaching one of the brown belts, I realized that I was able to automatically respond to a wrist grab and put him on the ground without even thinking about it.

Changing your thinking is learning a new skill in the same way you learn a physical skill. You don't expect to be a proficient golfer without swinging the club a few thousand times. In fact, in the book "Outliers" by

Malcolm Gladwell, he sets the magic number for success at 10,000 hours. According to Gladwell, when we examine the lives of the most successful people in any sport or profession, we typically find that they have spent at least 10,000 hours honing their craft.

Certainly, that doesn't mean that it will take 10,000 hours to change your thinking and to have a positive impact on your life. In fact, with frequent repetition of the rational challenging thoughts you should notice changes in four to eight weeks. However, you will need more repetition than that for the new way of thinking to become automatic.

Think of it this way. How many hours have you spent in your lifetime repeating the detrimental irrational self-talk? You likely have become very automatic in that way of thinking so now you need to re-train your brain to think in a different way. To do that you need to identify specific thoughts that are beneficial and to deliberately think those thoughts repeatedly.

Sometimes it is difficult to achieve the necessary repetition especially when you are first learning. That is why reading articles about CBT or motivational thinking can be helpful. The more you read, the more repetition you obtain. Also, that is why I have created some audios that focus on changing thinking. The audios provide repetition without a great deal of effort.

ERROR 2: Making Assumptions about CBT Based on Social Comparision.

Many times when I first explain cognitive therapy to clients and we examine their irrational thinking and explain why it is irrational, they exclaim "That makes so much sense! I've never thought about it that way before!"

However, for some people, once they understand the process of using irrational thought challenges, they believe that that they should just be able to think that way without having to examine their thinking and deliberately change it. They make the mistake of using social comparison assumptions. In particular, they might state "I should be able to just think this way. Everyone else doesn't have to do this (referring to the CBT methods)."

What they don't realize is their assumptions about others not using CBT techniques is inaccurate. Just about anyone who is successful in life is using these techniques! However, they may not have been taught the techniques of CBT specifically and they may not know the terms such as "mind-reading" or "generalizing" but they are still using the methods.

I tend to think that CBT is just a compendium of methods that have helped people achieve success for thousands of years. For instance, if you read the "Tao te Ching" examining it for cognitive concepts, you will find that it contains all the same challenges to irrational thinking and methods of successful living that you will find in any book by Aaron Beck, Albert Ellis, or David Burns.

If you listen closely to people, you will begin to recognize how they apply these methods in their lives. For instance, I frequently hear people (outside of the office) say when describing a difficult event, "...but then I told myself..." In other words, they were using self-talk to cope with a situation. I've also heard people say things such as "Then I took a deep breath and..." describing how they calmed themselves before tackling a problem.

So don't assume that because you are having to learn these methods and deliberately apply them that you are doing something others don't have to do. The only difference is that you may not have learned the methods naturally while growing up and you are learning them now. You may not have had the role models or opportunities to learn these methods when you were young.

People aren't naturally successful. Anyone who is successful, whether in life, sports, career, or relationships, has learned these methods at some point in their lives and are applying them routinely.

ERROR 3: Not Making CBT Methods a Lifestyle Change

Many people don't realize that CBT is training the brain to react differently to situations. As a result, they often mistakenly believe that recognizing the error in thinking and correcting it a few times will impact a change in their life. After they have some initial benefit they may believe that it isn't necessary to continue practicing the methods.

However, cognitive-behavioral changes need to be thought of as lifestyle changes just as exercise or an effective diet is a lifestyle change. You wouldn't expect to exercise for a few months, tone your muscles, stop exercising and expect your muscles to remain toned. Nor, is it reasonable to expect to lose weight by restricting your calories but not expect to gain weight when you return to eating everything you want.

A lifestyle change refers to any activity or behavior in which we engage regularly to maintain our health. In the case of the CBT methods, we make a change to improve our emotional fitness. But we can't expect that fitness or change to remain intact if we don't devote attention to practice.

People have come to realize that the body needs maintenance to retain fitness and optimum health. Too often, however, people think that the brain does not require maintenance. Therefore, many people obtain initial benefits from CBT but will revert to prior thinking patterns and behaviors due to the lack of attention to the CBT methods.

A good example of what may occur is the use of relaxation methods during a time of stress. The regular use of relaxation reduces the degree of stress experienced. However, many people are likely to think "I feel good. I don't need to do this relaxation anymore." However, stopping the relaxation practice may result in a return of the symptoms associated with stress.

Unfortunately, when this occurs some people may conclude "That relaxation stuff really doesn't work!" However, that is like saying exercise doesn't work because when you stop exercise you lose muscle strength.

ERROR 4: Not Using Relaxation Regularly

A closely associated error is that many people believe they should obtain benefits from relaxation even though they use it infrequently. I can't tell you how often clients report being stressed or anxious, yet when I ask them how often they engaged in relaxation, they respond "once" or "not at all." The same clients, however, may report a very positive response from the relaxation when they use it.

Again, the regular practice of relaxation trains the brain and the body to respond differently. This practice can create a very powerful response. You may be familiar with strong automatic negative responses. For instance, if you have been very stressed at work, you may notice your body reacting with symptoms when you drive by you place of employment on your day off. Our bodies can become conditioned to stress responses and associate the stress response with a place, person, or thing.

However, our bodies can also become conditioned to the relaxation response. I had an experience once that illustrated to me just how powerful the relaxation response can be. Years ago I listened to a particular tape (before CDs and MP3s) whenever I had a stress headache. One day, years later, while I was driving and listening to the radio, my entire body suddenly relaxed. I felt an incredible, soothing peacefulness. Wondering why I suddenly felt so relaxed, I realized that one of the songs that I used to listen to was playing on the radio. My body had become conditioned to relax to that music.

To be able to experience the powerful benefits of relaxation, however, it is necessary to practice regularly.

ERROR 5: Expecting Results Without Practice.

Very closely related to the importance of using the methods regularly, is the expectation many people have that using the methods one time should be effective. In the course of my practice, I have frequently heard a client say, "I tried breathing when I was anxious and it didn't work." When I asked them how much they had practiced prior to using it for the anxiety, they have told me not at all or very little.

If you were playing baseball, would you expect to hit the ball during a game if you had never tried hitting a ball before? Of course not! However, people often believe that the techniques of CBT should be effective when they have never practiced them. I think this may be due to the techniques being very similar to other normal daily behaviors so people think that they should just be able to do it without an previous training or practice.

For example, they might think that since breathing is something they have done every day of their lives, they should be able to just slow it down at will. However, the fallacy here is that although they have been breathing their entire lives, they have not been regulating their breathing at will. These are two different things.

To put this in perspective, if I asked you to lower your body temperature (without any artificial means such as ice) you might tell me I'm crazy, that it can't be done. But your brain has regulated your body temperature all your life just as it has regulated your breathing. So why do you believe that you can't lower your body temperature? Probably because you are not as aware of it as you are of your breathing. However, it is possible to change our physiological responses such as body temperature or blood pressure but it requires advanced practice.

The point is that practice is necessary to achieve the ability to change our physiological responses. We can't expect to be proficient at calming ourselves if we never practice the techniques. In CBT, homework is everything! (I know I repeated this concept several times but it's really important!)

ERROR 6: Believing CBT is the Same as Positive Thinking.

Another common error that people make about CBT is believing that it is all about positive thinking. I have frequently heard the remark, "I've tried that positive thinking—it doesn't work." However, CBT is not about positive thinking at all. In fact, positive thinking can be just as irrational and problematic as negative thinking.

For example, if someone believes "Everything will be okay. I don't need to worry" when a catastrophe is occurring, they may not take the necessary steps to control the situation which allows it to become worse. There are times when we need to be concerned so that we can take appropriate action.

However, cognitive therapy approaches this in a realistic manner. If we are worrying about something that is not likely to occur, then we are making ourselves feel bad about something for little reason. Therefore, CBT is practical and realistic. It is about assessing our thinking so that we can take the most reasonable approach.

ERROR 7: Believing That Emotions Are Always Irrational.

Sometimes people take CBT too far and believe "If I'm completely rational, I won't feel anything." In fact, I heard Albert Ellis, one of the founding fathers of cognitive therapy, state this idea at a seminar. He said that if we are always rational, we would never be angry.

I hope I misunderstood his statement because I disagree with this concept. We are emotional beings. Emotions are an important part of our information processing system. Emotions allow us to be aware of problems that need to be addressed. So I neither believe that normal human beings can be completely free of emotions nor do I believe that it is healthy for us to not have emotions. Without emotions we can miss very important information in our environment.

For example, if you are threatened on a dark street late at night, what do you think reacts first? Do you think your emotions react first warning you of danger? Or do you intellectually assess the situation first? The initial emotional reaction prepares us to flee or to fight. The intellectual response is to determine what might be the better decision.

Certainly, our emotions can cause us to over-react. For instance, on that dark street you might feel threatened when you really aren't. And that is where CBT can help. It helps us to learn to keep our emotions in perspective and to recognize when we may be over-reacting. However, it should not be used to prevent emotions completely.

Many CBT therapists are finding standard CBT too limiting in this regard and there has been much movement to mindfulness-based CBT, often referred to as MBCT. MBCT focuses more on learning to tolerate the various emotional states through mindfulness approaches rather than trying to eliminate them. The intellectualized approach in standard CBT can be problematic for people who already have a more analytical style with avoidance of emotions.

ERROR 8: Placing Demands on Rationality

As people learn about CBT, sometimes they come to believe that if being rational is good and benefits their life, they might place a demand on themselves: "I SHOULD be completely rational all of the time."

Now, if you know anything about cognitive therapy, you know that this statement is irrational in itself for several reasons. One, is that it is a perfectionistic demand. Such demands are impossible to achieve which then causes additional distress and dysfunction. Cognitive therapy is about decreasing stress by being realistic in our expectations of ourselves.

Another reason this statement is irrational is because it is black-and-white reasoning which is the idea that if something is bad, the opposite should be good. But this isn't the case either. CBT is about finding balance whereas black-and-white thinking is about extremes. Even CBT when done to the extreme can be unhealthy.

I have often had this issue with my anxious clients. When we work on exposures to anxiety-provoking situations so that they can desensitize to the situation, they may come to believe that it is always good to challenge anxiety. In other words, if you are afraid of something, you need to face it. But that is not true because sometimes challenging a fear unnecessarily can cause more harm than good.

In particular, this attitude can cause unnecessary stress which may interfere with someone challenging the fears that are important to face. A person needs to evaluate a fear, determine in what way and how much it affects satisfaction in life, and then decide whether it needs to be dealt with. If I have a fear of snakes that doesn't really affect my life much, why should I face it? However, if it prevents me from doing something that is very important to me such as riding my bike on a wooded path, then I might choose to face it.

ERROR 9: Using CBT to Justify Not Being Responsible for Change.

Another area where people need to be cautious with CBT is using it as a justification to not change or try. Although CBT teaches a person to appreciate all aspects of themselves, it is meant to be used for building a foundation that provides a stable base from which a person can make changes.

Practitioners of CBT believe that if you feel good about yourself, you will have greater confidence to try new things and to make improvements. You are more able to acknowledge your flaws or areas in need of improvement without that acknowledgment causing you to feel bad. So we teach acceptance of yourself.

However, some people may use this concept as justification for not taking responsibility to make changes: "I am fine just the way I am." Or, they may focus on only a very specific aspect of a cognitive belief without considering the complexity of the situation. For instance, if I tell a woman "Certainly, you have the right to expect your husband to help with half the housework, but all rights come with consequences and you also need to consider the consequences of that demand and determine how it affects your marriage" I might soon hear from her husband saying that his wife told him that he should help with half the housework! She ignored the rest of the challenge to her thinking.

My point here is that CBT shouldn't be used as a justification for behavior that is unhealthy or inappropriate. It is meant to help a person examine thoughts and choices so as to improve satisfaction with life and with relationships. It is not meant to help a person to continue problematic behaviors. So even if the words are the same, it may not always mean the same thing. You need to look at the underlying meaning as well. In other words, a "should" is not always a "should." And to paraphrase Freud, "Sometimes a should is just a should (and nothing else)."

ERROR 10: Placing Demands on Mindfulness

Another frequent error is the tendency to place expectations or demands on the practice of mindfulness. The problem with this is that demands prevent the experience of mindfulness so the more we try to be mindful the less we are able to be mindful.

I experienced this years ago as a college student. During a womens' retreat, we were instructed to listen to some music to allow us to experience imagery to develop personal understanding. As I am lying on the mat and trying to allow the music to create images, I became very frustrated. I thought "I'm not having any images! Nothing is occurring! Everyone else is going to experience something and share what they are experiencing. And I have nothing! How embarrassing! This is a waste of time."

As you can see, I was not mindful at all. Of course no images occurred. I was so caught up in my (irrational) thinking that I couldn't be mindful. Finally, in frustration, I gave up. I said to myself "I don't care! I'm just going to listen to the music and not worry about trying to have any images."

As I continued to listen to the music and let all my thoughts drift away, I suddenly had one of the most powerful images that has guided my life to this day. More importantly, I learned the critical concept of not trying to be mindful. Instead of trying to be mindful, learn to allow yourself to be mindful. The more you remove the demand of mindfulness, the more you will achieve it.

Things You Should Know About Cognitive Behavioral Therapy

You've probably heard about cognitive behavioral therapy (CBT), a method of treatment based on evidence focused on changing negative thoughts and behaviours. Almost every self-help article online seems to mention it: Sleep problems? Try CBT. Trauma from childhood? Maybe CBT can help. Anxiety, depression, low self-esteem, flying fear, hangnails? CBT is your answer.

Basically, you have a good chance of receiving CBT or knowing someone you have. So what's it about? Is it really relieving psychological distress, and if so, how? How much does it cost, and can you use the techniques alone? Such details can be a mystery to the public at large. Fortunately, I'm a clinical psychologist who uses CBT in my practice, so I should be able to answer most of your questions. Now seeing the most common questions, we will do a repetition of what we have seen throughout the whole book. Let's see if you remember all the topics

1. First, what's the CBT heck?

CBT is one of a number of psychotherapy treatment methods. It is based on the assumption that many of the problems in life stem from faulty thoughts and behaviors (from which "cognitive "comes). We can alleviate distress by deliberately shifting them towards healthier, more productive goals. In practice, cognitive behavioral therapy usually involves identifying and replacing problem thoughts and behaviors with healthier responses. For example, in social situations, Jane Doe is anxious and has begun to avoid gatherings in favor of isolating evenings at home. A CBT therapist can teach her about the irrationally triggered fear response, teach her how to shift her thoughts and relax her body, and develop an action plan to help her stay calm while participating in this weekend's party. They will evaluate what worked and what didn't work next week and tweak their methods until Jane can socialize comfortably.

2. What kind of problems can CBT help and how do I know if it's right for me?

CBT is used for everything from phobia, anxiety, depression, trauma, self-esteem and ADHD to relationship problems such as poor communication or your partner's unrealistic expectations. In essence, if it's a problem involving thoughts and behaviors (which covers a lot of ground), CBT has a treatment approach. Is it okay for you? That's a hard question. Do your problems have to do with how you think and act? Do you ruminate about a previous breakup, for example, or do you find yourself mindlessly shopping online? If so, yes, you might benefit from CBT. If you are more concerned about your purpose or meaning in life, or what moments of your past color you are today, there may be other approaches that are better suited to you (and we will get to that in question #9).

3. What is the popularity of CBT?

One of the reasons why CBT is so well known and widely used is that it has been so extensively studied. Studying is a good way because it emphasizes short, direct, solution-oriented interventions. In other words, the objective is to bring about clear, measurable changes in thinking and behaviour, which is a gold mine for researchers. It also means that you can see quick results. Since a high percentage of people in our practice deal with some form of anxiety (social anxiety, anxiety about health or disease, OCD, panic, etc.), it is a central part of the work to be able to gently challenge people to confront their fears and develop new ways of relating to their own thoughts " CBT provides us with the tools to encourage people to do something very unpleasant: confront the things they avoided"

4. What's going on in a CBT session?

CBT is a form of psychotherapy, so you can expect the early sessions to be what you would see in any initial therapy sessions: Discussing payment information and cancelation policy, your therapy goals, your history and your problems review. After that, you'll talk about your struggles and try to formulate together the most effective response. In essence, the customer brings the problems they want to overcome or the situations they find stressful, and the therapist and client work together to create an action plan. An action plan means identifying problem thoughts or behaviours, finding a way to change them and developing a strategy for this change in the coming week. There's "homework "here.

5. How are CBT homework?

CBT focuses on providing rapid (8 to 12 sessions, which are rapid according to therapeutic standards) and effective symptom reduction, which is best done throughout the week, not just during the therapy session. Typical homework may include relaxation exercises, keeping a journal of thoughts and emotions throughout the week, using worksheets that target a particular area of growth, reading a book that addresses your problems, or looking for situations to apply your new approach. For example, Jane may want to keep an eye on meet-up events that challenge her to overcome her fears while applying her new techniques for relaxation. Another example: Let's say that his negative internal self-talk is a major factor in John Doe 's depression-he constantly lowers himself on a loop. John and his CBT therapist can talk about a technique called "stop thinking, "in which he abruptly disrupts the flow of negative thoughts by yelling (in his mind) "stop! " as he turns his thoughts to something more positive, such as an affirmation or a meditation app. Homework can involve at least once every day practicing this technique until the next session. In the next session, John and his therapist will debrief, assess what worked and what didn't work, and tweak the process for the next week.

6. How long does CBT usually last?

One of CBT 's highlights is that it focuses on eliminating symptoms as quickly as possible, usually in a couple of weeks or months. Of course, people rarely have only one problem to work on in therapy, so this length depends on the number and severity of the problems, but briefness is essential for this approach. This leads to one of the main differences between CBT and many other forms of treatment. One of CBT 's founders, Donald Meichenbaum, says, "[We ask] what questions and how. While other treatment approaches spend a lot of time digging deeply and asking why you feel depressed, anxious or low self-esteem, CBT adheres to current thoughts and behaviors. CBT focuses on helping you reduce your fear rather than examining why you are afraid of snakes. While some people are happy to reduce their symptoms, others want to know why they first exist. Deeper approaches such as psychodynamic therapy may be more satisfactory for them.

7. Can people use CBT techniques outside of actual therapy sessions?

Did you ever keep a journal of gratitude? What about your donut intake monitoring? Did you track your daily steps or check your sleep? Then you're using some of CBT 's principles in your daily life. Many others CBT techniques can be found in books such as David Burns ' Feeling Good or Edmund Bourne 's Anxiety and Phobia Workbook, online, or in popular applications such as Headspace and Happify. But a period of time in structured therapy is still the best approach for a CBT course tailored to you and your problems.

8. How much does CBT cost and is it covered by insurance?

CBT is psychotherapy, so if your insurance covers psychotherapy or behavioral medicine, most, if not all, of your CBT treatment should be covered. If you pay out of the pocket, the cost of CBT ranges from free or sliding in some community clinics to $200 + per session in a private practice. Once again, the time someone spends on treatment is usually less than other treatment approaches, so it can be cheaper in the long run. On a therapist finder website such as Psychology Today or Good Therapy, you can search for a therapist who practices CBT and fits your budget.

9. Are there any sides to choose CBT compared to another type of treatment?

Some clients may feel that therapy is a place where they come and process their experiences with their therapist's gentle facilitation. Their main objective may not be to deal with a particular symptom or problem habit, but rather with general growth and a long-term relationship with a therapist. They may want to explore their memories, dreams and early relationships with their therapist's guidance. Since CBT can be a more direct and practical therapy style, it may not be helpful for someone who is looking for this kind of profound, relationship work. However, many qualified CBT therapists are very flexible in their approach and can adapt to the needs of a variety of clients. As Dr. Hsia admits, CBT is not without its criticisms. " CBT 's fair criticism highlights its assumptions about what helps people get better, "he says. Again, CBT focuses on symptoms rather than the deeper roots of these symptoms, and some psychologists who feel that the deeper roots are essential would consider CBT to be short-sighted. In the end, you have to find out what works best for you, which may take some trial and mistake. It may be helpful to talk to your therapist (or potential therapist) about what you are looking for and ask them how they approach the treatment. Whether you are receiving CBT treatment or another method, the most important thing is that you feel a safe, confident relationship with your therapist and that the treatment makes sense for you.

Pros & Cons of CBT Therapy

There is always a risk that bad feelings will return, but it should be easier for you to control them with your CBT skills. This is why it is important to continue to practice your CBT skills even after you feel better and have completed your sessions. CBT may not, however, be successful or appropriate for everyone.

Below are some of the advantages and disadvantages of the approach.

The benefits of CBT

In cases where the patient does not see an improvement with medication alone, cognitive behavioral therapy is effective. It does not take as long as other types of treatment. Cognitive behavioral therapy is also structured in such a way that it can be offered in different formats, including group therapy, self-help books and online programmes. Most people receive cognitive behavioral treatment because it teaches them practical strategies that can be applied to everyday life. It offers patients solutions to improve their minds every day.

Below are some specific benefits of CBT;

- can be as effective as medication in the treatment of certain mental health disorders and

- can be helpful in cases where medication alone has not worked. Compared to other speech therapies, it can be completed in a relatively short time.

- Focus on retraining your thoughts and changing behaviors to change your feelings. CBT 's highly structured nature means that it can be provided in various formats, including in groups, self-help books and computer programs.

- The skills you learn in CBT are useful, practical and helpful strategies that can be incorporated into everyday life to help you better cope with future stresses and difficulties even after treatment is complete.

CBT 's disadvantages

First of all, a high level of commitment is required for cognitive behavioral therapy. If you don't cooperate, your therapist can't help you. If you have a busy lifestyle, the frequency of cognitive behavioral therapy sessions may be a challenge. In addition, it may not be a good choice for people with complex mental problems or lcarning difficulties due to the highly structured nature of cognitive behavioral therapy. Many people fear cognitive behavioral treatment because it forces them to face their root of anxiety.

You should be prepared for an initial period of fear and discomfort if you choose this type of therapy. Another significant disadvantage is that cognitive behavioral therapy does not address other problems, such as families and social factors, that can have a significant impact on the patient. It is not suitable for everyone, however effective cognitive behavioral therapy is. Although it can work wonderful things for your friends, you may not benefit from it. In addition, cognitive behavioral therapy tends to focus instead of the cause on the specific problem. The therapy focuses on the identification of current problems rather than past problems. Someone with a mental condition resulting from childhood trauma, for example, may not see much improvement in cognitive behavioral therapy.

Below are some specific disadvantages of CBT;

• To take advantage of CBT, you must commit to the process. A therapist can help and advise you, but without your cooperation, you can't get rid of your problems.

• It can take a lot of your time to attend regular CBT sessions and perform any additional work between sessions.

• It may not be appropriate for people with more complex mental health needs or learning difficulties due to the structured nature of CBT.

• Since CBT can confront your emotions and anxieties, you may experience initial periods of anxiety or emotional discomfort.

• Some critics argue that CBT does not address the possible underlying causes of mental health conditions, such as an unhappy childhood, because it only addresses current problems and focuses on specific issues.

• CBT focuses on the ability of the individual to change himself (his thoughts, feelings and behaviors) and does not address broader problems in systems or families that often have a significant impact on the health and well-being of an individual.

CPSIA information can be obtained
at www.ICGtesting.com
Printed in the USA
BVHW091006040221
599247BV00013B/2234